# RECONNECTING RIVERSIDE with its RIVER:
## Integrating Historical and Urban Ecology for a Healthier Future

OCTOBER 2023

PREPARED FOR
CALIFORNIA STATE COASTAL CONSERVANCY

PROJECT DIRECTION
Sean Baumgarten
Kelly Iknayan
Robin Grossinger

AUTHORS
Sean Baumgarten
Lauren Stoneburner
Jennifer Symonds
Kelly Iknayan
Bronwen Stanford
Matthew Benjamin
Erik Ndayishimiye
Vanessa Lee

DESIGN AND PRODUCTION
Ruth Askevold
Jennifer Symonds

ILLUSTRATIONS
Vanessa Lee

PREPARED BY San Francisco Estuary Institute

IN COOPERATION WITH AND FUNDED BY
CALIFORNIA STATE COASTAL CONSERVANCY

SFEI
PUBLICATION #1133

### SUGGESTED CITATION

San Francisco Estuary Institute. 2023 Reconnecting Riverside with its River: Integrating Historical and Urban Ecology for a Healthier Future. Funded by the California State Coastal Conservancy. SFEI Publication #1133, San Francisco Estuary Institute, Richmond, CA.

### REPORT AVAILABILITY

Report is available online at sfei.org/projects/reconnecting-riverside-with-its-river

### IMAGE PERMISSION

Permissions rights for images used in this publication have been specifically acquired for one-time use in this publication only. Further use or reproduction is prohibited without express written permission from the individual or institution credited. For permissions and reproductions inquiries, please contact the responsible source directly.

### COVER IMAGE CREDITS

**(front cover)** (top) View from Mount Rubidoux, 1908. (Courtesy of the Library of Congress); (bottom) Sunset from Mount Rubidoux, 2020. (Photo by John Ko, courtesy of Unsplash); (middle circle) Re-imagined Santa Ana Creek walkways and access. (Illustration by Vanessa Lee, SFEI)

**(back cover)** (top to bottom) Riparian scrub and forest habitat at Martha McLean Anza Narrows Park. (Photo by SFEI); Mid-19th century map of Rancho Jurupa. (USDC 1854-58); Map of plant assemblages along the river (mapping by SFEI; data from Aerial Information Systems, Inc. 2012); An artistic depiction of a potential restored riparian corridor. (Illustration by Vanessa Lee, SFEI).

# CONTENTS

## 1 INTRODUCTION — ii
- The study area and regional context — 5
- Associated regional plans and projects — 6
- Acknowledgements — 8

## 2 HISTORICAL ECOLOGY — 10
- Methods and process — 11
- Findings — 16

## 3 LANDSCAPE CHANGE and CURRENT CONDITIONS — 24
- Land cover change — 26
- Change to the Santa Ana River — 32
- Urban ecology assessment — 36

## 4 OPPORTUNITIES and STRATEGIES — 50
- Introduction — 51
- General recommendations — 52
- Santa Ana River riparian corridor — 54
- Arroyos — 58
- Wetland habitats — 60
- Riversidean sage scrub and forbland — 62
- Urban forest — 66
- Conclusion — 67

## REFERENCES CITED — 68

## APPENDIX A: Detailed Methods — 72

## APPENDIX B: Plant Palettes — 76

# 1 INTRODUCTION

Brittlebush *(Encelia farinosa)*. (Photo by Gentry George, courtesy CC 4.0)

In urban areas throughout the United States, residents and planners are seeking to enhance the resilience of their communities to climate change and other stressors. There is a growing recognition that many common urban forms of the 20th century, from lawns to flood control channels to strip malls, provide a narrow range of benefits and do not maximize the potential of the landscape either in supporting native biodiversity or providing ecosystems services to people. Climate change impacts are projected to exacerbate stresses on both human communities and native biodiversity, through more extreme heat, more frequent periods of drought, and more frequent and severe storms accompanied by increased flooding. Residents and planners are increasingly looking to nature-based solutions to meet the challenges of the 21st century, from urban heat island effects, to habitat loss and fragmentation, to spread of invasive species, to water scarcity.

The convergence of urban areas and large river systems presents both unique challenges and opportunities. Dams and engineered levees, while protecting developed areas from flooding, have largely failed in providing other diverse benefits of a healthy river system, such as riparian wildlife habitat, groundwater recharge, and access for recreation and other cultural uses. Urban rivers, once viewed primarily as a water supply to be controlled and manipulated, are increasingly valued for the multiple benefits they can provide to both people and wildlife. Cities around the world are restoring their waterfronts to be vibrant features of the urban landscape.

In Southern California, the Santa Ana River is one such urban river with potential to provide a wide assortment of benefits to millions of Californians. The river flows nearly 100 mi (160 km) from its headwaters in the San Bernardino Mountains to its mouth near Newport Beach, Orange County. Being the largest watershed in the region, the river and its tributaries support a huge diversity of habitats and species. The Santa Ana River Conservancy Program, a program of the State Coastal Conservancy (SCC), was established by the California State Legislature in 2014 to sustain the vital natural and cultural resources of the Santa Ana River, focusing on the Santa Ana River Parkway encompassing lands within 0.5 mi (~0.8 km) of the river mainstem. The program supports a range of resource management goals including wildlife habitat protection and restoration, preservation of agricultural land, public access to the river for recreational and educational benefits, creation of a continuous Santa Ana River Trail system, and natural flood conveyance and water quality maintenance (Placeworks 2018).

In the City of Riverside, local planners and communities, working with SCC, have identified a need for science-based design guidance to inform habitat restoration and other resource management efforts within the Santa Ana River Parkway and surrounding areas. For instance, the Riverside Gateway Parks project led by the City with support from SCC, is in the process of designing a series of open spaces, habitat areas, and recreational facilities within the Parkway. The Santa Ana River is one of the most recognizable landmarks in Riverside (which derives its name from the river), and thus protecting and enhancing the connection between the river and urban areas is important not only for local biodiversity but also for the City's very identity.

## PROJECT GOAL

The goal of this project is to provide science-based design guidance for optimizing restoration planning for a portion of the Santa Ana River Parkway in and around the City of Riverside, using information derived from historical ecology and urban ecology research.

The goal of this project is to provide science-based design guidance for optimizing restoration planning for a portion of the Santa Ana River Parkway in and around the City of Riverside, using information derived from historical ecology and urban ecology research. This report synthesizes information from both historical and present-day landscape analyses to develop recommendations that support ecological processes and meet present-day species needs. The specific objectives of this project are to:

- Develop a map and associated information documenting historical landscape patterns and processes within the study area;
- Conduct spatial analysis characterizing landscape change over time;
- Assess current landscape patterns and ecological support functions provided by the river and surrounding urban areas; and
- Develop guidance and recommendations for restoration and open space design.

This current effort examines a small portion of the Santa Ana River Parkway and its surroundings within and around the City of Riverside, including a number of Gateway Parks. The project is envisioned as a pilot study that will be expanded to other portions of the Santa Ana River Parkway and the broader watershed in the future. The report does not represent a comprehensive management plan, and we do not make specific recommendations for land acquisition or conservation. Instead, the multi-benefit strategies and recommendations presented here are intended to inspire and guide planning efforts by a range of interest groups, including City planners, tribal communities, restoration practitioners, and the general public.

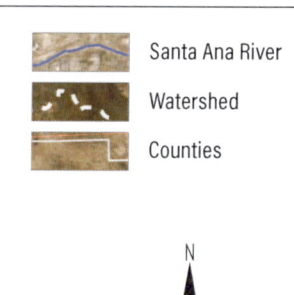

**Figure 1.1. (right) Map of the study area's position within the Santa Ana River watershed and key regional landmarks.** The study area covers the Santa Ana River through the City of Riverside and the surrounding lands to a distance of approximately 2.5 mi (4 km) on either side of the river corridor.

# The study area and regional context

The study area centers around the Santa Ana River where it runs through portions of the cities of Riverside and Jurupa Valley (Figs. 1.1, 1.2). It lies near the geographic center of the Santa Ana River watershed, a nearly 2,500 sq mi (~6,400 sq km) area that drains from multiple mountain ranges in California's Transverse and Peninsular Ranges and reaches the Pacific Ocean near Newport Beach. On its course, the river passes through the broad Riverside lowlands bioregion, past a narrow bottleneck at Santa Ana Canyon, and through various cities west of the Santa Ana Mountains. Over the course of the last century, extensive modifications have occurred both to the river's hydrology and the watershed's land cover, diminishing the area's ability to support native plants and wildlife.

The upper reaches of the Santa Ana River watershed support relatively intact ecosystems with little urbanization, including large areas of protected lands in the San Gabriel, San Bernardino, and San Jacinto mountains. These mountain ranges support diverse ecosystems, including coastal sage scrub, montane chaparral, coniferous forests, and broad-leaved forests. Various species of conservation concern reside in these habitat areas, such as mountain yellow-legged frog (*Rana muscosa*), Santa Ana speckled dace (*Rhinichthys osculus*), and California spotted owl (*Strix occidentalis occidentalis*) (Western Riverside County Regional Conservation Authority 2003, ICF 2020).

The Riverside lowlands bioregion, which contains the study area, has experienced significantly more habitat loss and urbanization than the surrounding mountains. This bioregion falls in the rain shadow of the Santa Ana Mountains and is characterized by a more arid climate than other areas of the watershed. Historically, the upland portion of this region largely contained sage scrub and forbland, while a mixture of riparian forest/scrub, alluvial scrub, alkali meadow, and river wash occurred along the Santa Ana River and its tributaries. Today, commercial, residential, and industrial land uses have supplanted most of the upland habitats and encroached upon riparian areas. Undeveloped habitat areas occur along the river or on rugged, elevated terrain, such as in the Jurupa Hills and Box Springs Mountains. Remnant upland areas support populations of coastal California gnatcatchers (*Polioptila californica californica*) and Bell's sparrow (*Artemisiospiza belli*), among other protected species, while riparian areas continue to provide habitat for sensitive species such as the least Bell's vireo (*Vireo bellii pusillus*) and aquatic species such as the Santa Ana sucker (*Catostomus santaanae*).

Expanding and enhancing habitat areas within the study area can help to better support the region's wildlife while providing more space for recreation and offering additional ecosystem services. The following sections outline the key components necessary to support biodiversity in this urbanized landscape, explain the current state of these elements in the study area, and opportunities to improve them in public and private spaces.

**Figure 1.2. Map of the study area, which encompasses nearly 13,000 ac (5,260 ha).** The study area extends to the northeast of Highway 60 and Fairmount Park, and it reaches southeast to Martha McLean Anza Narrows Park.

## Associated regional plans and projects

Several existing plans and projects guide restoration and development within the study area. The City of Riverside leads parkway improvement projects, eight of which fall within the study area boundaries and include: Fairmount Park/Camp Evans, Carlson Park and St. Francis Falls, Loring Park, Santa Ana River Greenway, Tequesquite North Extension, 5200 Tequesquite Ave, Tequesquite South Extension, and Martha McLean Anza Narrows Park (Fig. 1.3).

In addition to the Santa Ana River Parkway and Open Space Plan (Placeworks 2018), there are many other plans that influence restoration and development within this report's area of interest. These include plans and strategies for water resource management, habitat restoration, recreation, and climate action:

**WATER RESOURCES MANAGEMENT**

- One Water One Watershed Integrated Regional Water Management Plan for the Santa Ana River Watershed (OWOW)
- Coastal Conservancy Strategic Plan
- California Water Action Plan

**HABITAT CONSERVATION/RESTORATION**

- Western Riverside County Multiple Species Habitat Conservation Plan
- Upper Santa Ana River Habitat Conservation Plan Comprehensive Adaptive Management and Monitoring Program
- Upper Santa Ana River Wash Habitat Conservation Plan
- California State Wildlife Action Plan

**RECREATION**

- Santa Ana River Corridor Trail System Master Plan
- City of Riverside Comprehensive Park, Recreation & Community Services Master Plan

**CLIMATE ACTION**

- City of Riverside Climate Action Plan
- City of Riverside Green Action Plan
- County of Riverside Climate Action Plan
- San Bernardino Valley Municipal Water District Climate Adaptation and Resilience Plan

These plans and programs work together to guide the resiliency of the Santa Ana River watershed and have been referenced in the development of this report.

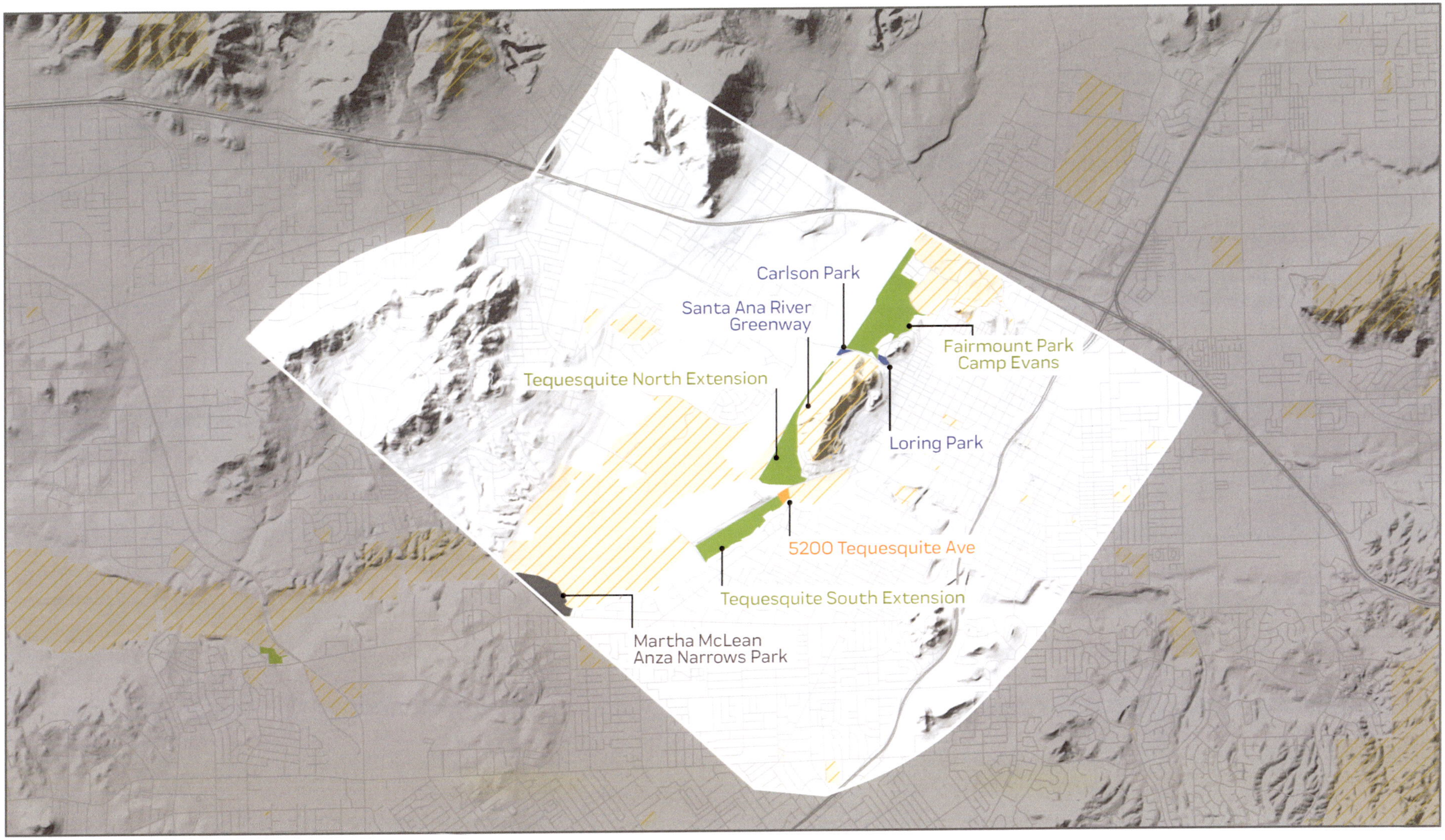

**Figure 1.3. Riverside Gateway Parks projects and protected open space along the Santa Ana River.** The projects create an almost continuous corridor along the southern bank of the Santa Ana River through the study area. Habitat quality and type are indicated by color.

**HABITAT QUALITY and TYPE**

- Mostly Non-Native Weedy
- Degraded Native
- Barren
- Developed Park/Grass
- Other Protected Open Space (California Protected Areas Database, GreenInfo Network 2020)

SANTA ANA RIVER HISTORICAL ECOLOGY

# Acknowledgements

This project was funded and made possible by the California State Coastal Conservancy (SCC). We would like to express our gratitude to the numerous SCC staff who participated in the project, including Greg Gauthier, Megan Cooper, Joel Gerwein, Rodrigo Garcia, and Danh Lai. We would like to extend special thanks to Greg and Megan for helping develop the vision for the project, and to Greg for his invaluable guidance and support throughout the course of the project.

We are deeply grateful to the members of our technical advisory committee for their guidance, technical advice, and enthusiastic contributions to the project: Dr. Loralee Larios (UC Riverside), Dr. Travis Longcore (UC Los Angeles), Dr. Bruce Orr (Stillwater Sciences), and Dr. Sophie Parker (The Nature Conservancy). We received additional helpful advice on best practices for coastal sage scrub restoration from Margot Griswold (Land IQ), Travis Brooks (Land IQ), Edith Allen (UC Riverside), and Sandy DeSimone (Audubon California Starr Ranch Sanctuary). Aaron Saubel (Malki Museum) generously shared Traditional Ecological Knowledge associated with plant species historically documented within the study area.

We are thankful for the input, feedback, and collaboration from our project partners: Alisa Sramala (City of Riverside), Randy McDaniel (City of Riverside), Pamela Galera (City of Riverside), Kai Palenscar (San Bernardino Valley Municipal Water District), Heather Dyer (San Bernardino Valley Municipal Water District), Chris Jones (San Bernardino Valley Municipal Water District), Joanna Gibson (San Bernardino Valley Municipal Water District), Patricia Lock Dawson (City of Riverside; PLD Consulting), Ian Achimore (Santa Ana Watershed Project Authority), Ray Hiemstra (OC CoastKeeper), Diana Ruiz (Riverside Corona Resource Conservation District), Ernesto Alvarado (Riverside Corona Resource Conservation District), Rachel Hamilton (Rivers and Lands Conservancy), Dana Rochat (The Wildlands Conservancy), Frazier Haney (The Wildlands Conservancy), Nick Deyo (ICF), Matt Romero (Studio MLA), Eden Ferry (Studio MLA), Megan Horn (Studio MLA), Jan Dyer (Studio MLA), and Zach Kantor-Anaya (The Wildlands Conservancy). Thank you to all of the staff at archives and source institutions visited and consulted throughout the course of the project.

A number of additional SFEI staff contributed to this project, including Stephanie Panlasigui, Scott Dusterhoff, Robin Grossinger, Cate Jaffe, Melissa Foley, Letitia Grenier, Erin Beller, and Samuel Safran. We also extend our thanks to David Ludeke, an intern from the Stanford University Bill Lane Center for the Environment, for assistance in creating the historical aerial photomosaic and other research contributions.

(top) Union Pacific Bridge across the Santa Ana River. (bottom) View from Mount Rubidoux, ca 1900s. (Photographs courtesy of the Museum of Riverside, Riverside California)

# 2
# HISTORICAL ECOLOGY

# Methods and process

## What is historical ecology?

Successful ecosystem restoration requires an understanding of the natural conditions that supported native species prior to recent landscape modifications. However, drastic land use and ecosystem changes over the last two centuries have made it challenging to understand the ecological patterns and processes that characterized the Santa Ana River and the Riverside region in the recent past.

Historical ecology is an interdisciplinary field that uses historical data to reconstruct the form and function of the past landscape, helping us better understand the contemporary landscape (and how it has changed over time) and envision future ecological potential. A deep understanding of the Riverside region's historical ecology yields foundational information about the historical distribution, composition, and structure of vegetation communities, the wildlife species that depended on them, the physical processes and controls that shaped landscape patterns, and a range of other topics. Information about natural landscape functioning reveals restoration opportunities and constraints in the present landscape, and is the basis for selecting locally appropriate restoration targets that will maximize benefits for native biodiversity. By incorporating our knowledge of the historical landscape, we can create a vision for how to best support a healthy, functioning ecosystem today and into the future.

## Data collection and compilation

We reconstructed historical landscape conditions within the study area by interpreting and synthesizing a large number of historical cartographic, textual, and photographic materials. Historical data were collected from a range of archives and online databases, including the California State Archives, California State Library, California Historical Society, Bancroft Library at UC Berkeley, Jepson Herbarium and Museum of Vertebrate Zoology at UC Berkeley, Riverside Public Works Department, San Bernardino County Surveyor, Claremont Colleges Digital Library, Loyola Marymount University, Library of Congress, Bureau of Land Management, University of Southern California Digital Library, Biodiversity Heritage Library, Consortium of California Herbaria (CCH), Western Foundation of Vertebrate Zoology (WFVZ), Vertnet, California Digital Newspaper Collection, David Rumsey Map Collection, and numerous others.

Key cartographic data sources included mid-19th century General Land Office (GLO) surveys, irrigation and flood control maps, parcel and subdivision maps, USDA soil surveys, USGS topographic maps, and Wieslander Vegetation Type Mapping (VTM) (Fig. 2.7 on page 23). Key textual data sources included 18th-century Spanish explorer accounts, mid-19th century Pacific Railroad Survey reports, early

> **KEY MESSAGE**
>
> Historical ecology is an interdisciplinary field that uses historical data to reconstruct the form and function of the past landscape, helping us better understand the contemporary landscape (and how it has changed over time) and envision future ecological potential.

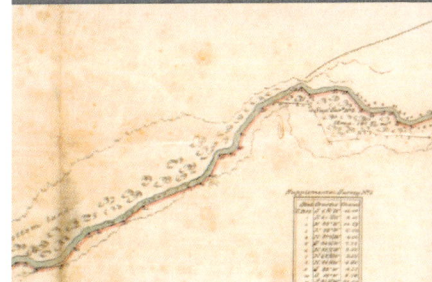

**Figure 2.1. "Village at Jurupa Rancho, base of Mt. Rubidoux**, near San Bernardino inhabited by Cahuilla, Serrano, and probaby some Gabrielino refugees. Photograph by C.C. Pierce, 1890" (Bean and Smith 1978). (PF20_858, courtesy of Huntington Library)

travelogues, GLO field notes, newspaper articles, and botanical and zoological specimen records and collector notes. Photographs included both early landscape photos and aerial imagery.

High-value spatial data sources were compiled into a geographic information system (GIS), so that they could be overlaid and synthesized with both historical and modern geophysical, climate, and land use data in order to compare across space and time. A subset of historical maps were georeferenced, and high-value textual excerpts (e.g., GLO field notes) and landscape photographs were geolocated. A photomosaic of the earliest available aerial imagery, from 1931, was created by orthorectifying high resolution scans in ERDAS IMAGINE (Fig. 2.2). Geospatial data were compiled in ArcMap 10.7.

## Data synthesis and mapping

Historical data were synthesized to develop a set of GIS layers representing average ecological conditions in the Indigenous landscape prior to major Euro-American modifications (ca. 1850). Land cover was classified as one of the following habitat type classes: Riversidean Sage Scrub/Forbland; Riparian Forest; Riparian Forest/Scrub; Alluvial Scrub, Alkali Meadow, and River Wash; Freshwater Emergent Marsh, Chaparral, and Vernal Pool Complex (Table 2.2). Several of these classes combine two or more habitat types (Riversidean Sage Scrub/Forbland; Riparian Forest/Scrub; Alluvial Scrub, Alkali Meadow and River Wash) where it was not possible to consistently differentiate them in the historical mapping. Stream channels were mapped as two-dimensional line features.

Feature boundaries and channel configurations were mapped from the most spatially accurate sources representative of pre-modification conditions. Key sources for mapping historical habitat types included the historical soils map (Nelson et al. 1915) and historical aerial photographs, and key sources for mapping historical channels included historical aerial photographs and USGS topographic maps. While mapping required us to draw sharp boundary lines, in reality the boundaries between habitat types would often have involved more gradual transitions.

Wherever possible, the classification and extent of each feature was verified using secondary sources. This verification through multiple independent data sources helped to uncover (and often resolve) inconsistencies between individual sources and reveal persistent landscape features and patterns. Rather than portray conditions at a specific point in time, we endeavored to map the general diversity and distribution of habitat types in the Indigenous landscape during average conditions just prior to significant Euro-American landscape modification.

Historical sources differ widely in terms of accuracy, level of detail, spatial extent, and bias. While no single source provides a complete picture of the historical landscape, the comparison and synthesis of multiple independent sources allows for a much more accurate reconstruction. Each feature was attributed in GIS with supporting sources and certainty levels representing our confidence in feature classification (interpretation), size, and location. Table 2.1 shows certainty levels and criteria used for mapping.

For more details about SFEI's general historical ecology methods, please refer to Grossinger et al. (2007), Beller et al. (2011), and Safran et al. (2017).

**Figure 2.2. 1931 aerial image of the Santa Ana River and floodplain just upstream of the Riverside Narrows.** A large, persistent riparian forest occupied this portion of the river corridor during the 19th and early 20th centuries. (Fairchild Aerial Surveys 1931)

Table 2.1. Definitions of certainty levels, which were used in the historical ecology mapping process to describe our confidence in each feature's interpretation, size, and location (Grossinger et al. 2007). Interpretation certainty describes our confidence that the habitat type assigned to the feature is accurate and that the feature is representative of the historical period. Size certainty describes our confidence that the feature's spatial extent is accurately depicted. Location certainty describes our confidence that the feature existed at the mapped location.

| Certainty level | Interpretation | Size | Location |
| --- | --- | --- | --- |
| High / "Definite" | Feature definitely present in the Indigenous landscape (before Euro-American modification) | Mapped feature expected to be 90% - 110% of actual feature size | Expected maximum horizontal displacement less than 150 ft (50 m) |
| Medium / "Probable" | Feature probably present in the Indigenous landscape (before Euro-American modification) | Mapped feature expected to be 50% - 200% of actual feature size | Expected maximum horizontal displacement less than 500 ft (150 m) |
| Low / "Possible" | Feature possibly present in the Indigenous landscape (before Euro-American modification) | Mapped feature expected to be 25% - 400% of actual feature size | Expected maximum horizontal displacement less than 1,600 ft (500 m) |

## Defining historical habitat types

Table 2.2. Table summarizing the key characteristics of habitat types historically present within the study area.

| Habitat Type | Description | Characteristic Plants | Physical Characteristics |
|---|---|---|---|
| **Riparian Forest** | A woody vegetation community dominated by deciduous trees, located adjacent to the Santa Ana River channel or on surrounding floodplain terraces. Areas mapped as riparian forest include some areas of sparsely vegetated channel bed and sandbars. | Fremont cottonwood (*Populus fremontii*), willow (*Salix* spp.; e.g., sandbar willow [*S. exigua*], Goodding's willow [*S. gooddingii*], and red willow [*S. laevigata*]), California sycamore (*Platanus racemosa*). | Supported by persistent high groundwater levels and seasonally variable streamflow. Forest regeneration requires scouring flood flows. |
| **Riparian Forest / Scrub** | A dynamic mosaic of riparian forest intermixed with lower-statured, early successional shrubs and small trees, shifting over time as a result of streamflow and flooding. Areas mapped as riparian forest/scrub include some areas of sparsely vegetated channel bed and sandbars. | Fremont cottonwood (*Populus fremontii*), willow (*Salix* spp.; e.g., sandbar willow [*S. exigua*], Goodding's willow [*S. gooddingii*], and red willow [*S. laevigata*]), California sycamore (*Platanus racemosa*), mulefat (*Baccharis salificolia*), willow baccharis (*Baccharis salicina*). | Supported by persistent high groundwater levels and seasonally variable streamflow. Forest regeneration requires scouring flood flows. |
| **Alluvial Scrub** | A relatively open, shrub-dominated community occurring on alluvial fans and floodplains. Alluvial scrub is dominated by drought-deciduous shrubs, but also includes a component of evergreen and riparian shrubs. Within the study area, often interspersed with river wash and alkali meadow. | Scale broom (*Lepidospatum squamatum*; indicator species), sugar bush (*Rhus ovata*), white sage (*Salvia apiana*), California buckwheat (*Eriogonum fasciculatum*), chaparral yucca (*Yucca whipplei*), California croton (*Croton californicus*), yerba santa (*Eriodictyon* spp.), Santa Ana River woollystar (*Eriastrum densifolium* ssp. *sanctorum*). | Occurs on alluvial fans and floodplains characterized by periodic scouring floods and coarse-textured, well-drained soils (Hanes et al. 1989). |
| **Alkali Meadow** | Seasonal wetlands characterized by moderately alkaline soils, seasonal flooding, and a salt-tolerant plant community. Within the study area, often interspersed with river wash and alluvial scrub. | Saltgrass (*Distichlis spicata*), yerba mansa (*Anemopsis californica*), marsh fleabane (*Pluchea odorata*), horned sea blite (*Suaeda calceoliformis*), seaside heliotrope (*Heliotropium curassavicum*), alkali buttercup (*Ranunculus cymbalaria*). | Restricted to zones of shallow groundwater (Elmore et al. 2006) subject to seasonal to intermittent flooding, with subsequent drying through the summer. Characterized by poorly drained, clay-rich, salt-affected soils. |

| Habitat Type | Description | Characteristic Plants | Physical Characteristics |
|---|---|---|---|
| River Wash | Sparsely vegetated community with exposed sand, scattered willows, and seasonal grass cover. Within the study area, often interspersed with alluvial scrub and alkali meadow. | N/A | Flooding and frequent disturbance. |
| Freshwater Emergent Marsh | Permanently flooded to intermittently exposed, permanently saturated palustrine wetland. Interspersed with sand bars, sparse trees, and pools dominated by aquatic vegetation. | Tules (*Schoenoplectus* spp.; e.g., Olney's bulrush [*S. americanus*]), rushes (*Juncus* spp.; e.g., frog rush [*J. ambiguus*]), cattails (e.g., broadleaf cattail [*Typha latifolia*]). | Supported by high groundwater and groundwater-fed springs, with frequent or semi-permanent flooding and permanently saturated soils. |
| Riversidean Sage Scrub | Shrubland dominated by drought-deciduous, soft-stemmed shrubs generally less than 6 ft (2 m) in height, intermixed with grasses and forbs. A subtype of coastal sage scrub characterized by a greater proportion of plants with desert affinities (Barbour et al. 2007). Within the study area often occurring in a mosaic with forbland. | California sagebrush (*Artemisia californica*), black sage (*Salvia mellifera*), California buckwheat (*Eriogonum fasciculatum*), California brittlebush (*Encelia farinosa*). | Variable topography; frequently occurs on steep hillslopes (especially south-facing) and on thin, rocky soils. |
| Forbland | An herbaceous plant community dominated by perennial and annual forbs and perennial bunchgrasses. Within the study area often occurring in a mosaic with Riversidean sage scrub. | California poppy (*Eschcholzia californica*), Kellogg's tarweed (*Deinandra kelloggii*), Indian paintbrush (*Castilleja exserta*), lupine (*Lupinus* spp.), dwarf checkerbloom (*Sidalcea malviflora*), Johnny-jump-up (*Viola pedunculata*), coastal tidytips (*Layia platyglossa*), foothill needlegrass (*Nasella lepida*), prairie junegrass (*Koeleria macrantha*). | Hillslopes and valley floors, often on relatively deep and fine-textured soils. |
| Vernal Pool Complex | Seasonally or intermittently flooded depressions, characterized by a relatively impermeable subsurface soil layer and distinctive vernal pool flora. | Spike rush (*Eleocharis* spp.), hairy water-clover (*Marsilea vestita*), neckweed (*Veronica peregrina*) | Topographic depressions with an impervious hardpan layer beneath shallow soil, causing seasonal flooding. |
| Chaparral | Shrubland dominated by chamise and other evergreen, sclerophyllous shrubs up to 13 ft (4 m) in height. | Chamise (*Adenostoma fasciculatum*), Eastwood manzanita (*Arctostaphylos glandulosa*), wild-lilac (*Ceanothus crassifolius*) | Xeric-to-mesic soils on hot, dry sites. |

# Findings

## *River corridor*

The Santa Ana River occupied a broad river corridor through this portion of the San Bernardino Valley historically. During the mid-19th century period represented by the historical mapping (Fig. 2.6), the mainstem river channel split into two branches approximately 0.5 mi (0.8 km) upstream of the present-day Mission Blvd bridge, which reconnected approximately 3 mi (4.8 km) downstream near present-day Martha McLean Anza Narrows Park. While the depiction in the historical mapping represents a single snapshot in time, the river corridor was a dynamic environment frequently changing in response to episodic flood events, which removed vegetation, filled old channel courses, and scoured new ones. Reconstructing historical hydrologic patterns was not the focus of this study, but the river generally was characterized by pronounced seasonal variability in streamflow and periodically experienced massive flood events (Hall 1888a).

The broad, sandy beds of the Santa Ana River channels (Williamson 1856) were bordered by a mix of **riparian forest** and **scrub** in varying stages of successional development, ranging from sparsely vegetated river wash to mature forest (Fig. 2.3). Mature riparian forests were dominated by an overstory of willow (*Salix* spp.), Fremont cottonwood (*Populus fremontii*), and California sycamore (*Platanus racemosa*), along with species such as black walnut (*Juglans californica*) and alder (*Alnus* spp.) (Tyson et al. 1851, Hancock 1858, Hayes 1863, Unknown 1872, Font and Brown 2011). Earlier successional riparian scrub was dominated by species such as desert willow (*Salix exigua*), mulefat (*Baccharis salicifolia*), and willow baccharis (*B. salicina*) (Kelly et al. 2005). The riparian understory supported a range of herbaceous plants such as California mugwort (*Artemisia douglasiana*), common evening primrose (*Oenothera elata*), western goldenrod (*Euthamia occidentalis*), desert wild grape (*Vitis girdiana*), hairy waterclover (*Marsilea vestita*), and catchfly prairie gentian (*Eustoma exaltatum*) (Reed 1916, data from CCH).

The river and adjacent riparian environments were critical resources for many species of wildlife in the otherwise arid environment, providing habitat for reptiles and amphibians such as western pond turtle (*Actinemys marmorata*), songbirds such as Bell's vireo and willow flycatcher (*Empidonax traillii*), waterfowl, and fish such as the now-federally endangered unarmored threespine stickleback (*Gasterosteus aculeatus williamsoni*) and the federally threatened Santa Ana sucker (Rutter 1896, Snyder 1908, data from WFVZ).

A large, persistent grove of riparian forest, measuring over 5,000 ft (~1,500 m) wide in some areas, existed just upstream of the Riverside Narrows on the southwest side of the study area. Supported by groundwater forced to surface by the geologic constriction at the River Narrows, as well as flow from Spring Brook and Tequesquite Arroyo upstream (Mendenhall 1905), this patch of forest was consistently documented by observers throughout the mid-19th and early 20th centuries (e.g., USDC 1854-58). A surveyor in 1872, for instance, described the vegetation in the area as a "dense growth of low willows and sycamores in low swampy bottom" (Unknown 1872). In contrast, the northern portion of the river supported a narrow corridor with a heterogeneous mix of riparian forest and scrub.

Further from the river channel, less frequently flooded portions of the Santa Ana River corridor, as well as the Tequesquite Arroyo canyon to the east, were dominated by **alluvial scrub**, with a smaller amount of **alkali meadow** and **river wash** (Fig. 2.4). Alluvial scrub supported a wide variety of drought-deciduous (i.e., shed leaves during the dry season) and evergreen shrubs and forbs such as California scale broom (*Lepidospartum squamatum*), white sage (*Salvia apiana*), sugar sumac (*Rhus ovata*), California buckwheat (*Eriogonum fasciculatum*), chaparral yucca (*Yucca whipplei*), California croton (*Croton californicus*), and yerba santa (*Eriodictyon* spp.) (data from CCH), and provided habitat for wildlife such as the now federally-endangered San Bernardino kangaroo rat (*Dipodomys merriami parvus*) (data from Vertnet).

Figure 2.3. Santa Ana River near Riverside, 1896. (A378-41, courtesy of the Museum of Riverside, Riverside, California)

> "The Santa Ana River is one with a good deal of water, and sunken very deep in… Upon its banks it nourishes a few cottonwoods, the only trees that are to be found on all of these plains."
>
> —Font and Brown [1775] 2011

**KEY MESSAGE**

The Santa Ana River historically occupied a broad river corridor bordered by a mix of riparian forest and scrub, alluvial scrub, alkali meadow, and river wash habitat types.

# NATIVE LAND MANAGEMENT

(top) Cahuilla woman making a basket, circa 1905 (photograph courtesy of CC 4.0). (bottom) Sumac (*Rhus trilobata*) in Southern California (photograph by Tom Benson, courtesy of CC 4.0).

Evidence for human presence in the Riverside region dates back to at least 9,000 years ago (Horne and McDougall 2007). At the time of European colonization, the Kizh and Tongva (Gabrielino), Payómkawichum (Luiseño), Cahuilla, and Serrano peoples lived within the region now occupied by the City of Riverside and surrounding areas (Milliken et al. 2010). This portion of the Santa Ana River watershed lies within a transitional zone between the traditional territories of these major tribal groups.

Though the ethnographic and archaeological data is incomplete, early records document several Indigenous villages in and around the study area (Fig. 2.1). For instance, while camped on the banks of the Santa Ana River near the Riverside Narrows in March 1774, Anza observed "a village… whose number would be more than sixty persons" (Bolton 1930). Horuuvnga, a Gabrielino community whose name reportedly derives from the Gabrielino word hurúuvar, meaning sagebrush (*Artemisia californica*), was located within the area just west of Riverside that later became the Jurupa land grant (McCawley 1996). Saubel and Elliott (2004) also describe a Cahuilla village known as Húlvel Pá' where "the road drops down to Riverside" near Mt. Rubidoux. Networks of trails connected villages as well as hunting and gathering sites (Bean 1972).

Tribes in the region practiced a hunter-gatherer lifestyle and utilized a wide range of natural resources for food, clothing, basketry material, construction material, medicine and other needs. Acorns were staple food resources, along with a wide variety of seeds, fruits, roots and tubers, greens, and fungi (Bean 1972, Bean and Saubel 1972, Bean and Smith 1978, Bean and Shipek 1978, McCawley 1996). (Though oak groves were not documented within the study area historically, acorns were likely gathered from surrounding areas.) Deer, antelope, small mammals, birds, and other animals were hunted for meat as well as fur, skins, and other implements (Bean and Smith 1978, Bean and Shipek 1978, McCawley 1996). Trade between villages and amongst tribal groups was common (Bean 1972, McCawley 1996).

Fire was used to promote the growth of favored plants, remove dead plant material, control pests and pathogens, aid in hunting animals, and for other purposes. For instance, the Gabrielino periodically burned grasslands (or forblands) to increase yields of plant foods (McCawley 1996). Bean (1972) reports that the Cahuilla burned grasslands/forblands to increase yields of chia *(Salvia columbariae)* seeds and control grasshopper and locust populations, and burned brush to flush game animals. The Serrano likewise used fire to increase yields of chia (Bean and Smith 1978). Bean and Shipek (1978) report that the Luiseño used fire for rabbit drives and to promote the growth of basket grasses, grass seed, yucca, and other useful plants; burning of crop plants was conducted every three years or less. The Luiseño also burned sumac (*Rhus trilobata*) to stimulate the growth of shoots for basketry (Anderson 2005). The intentional use of fire was likely an important factor in shaping vegetation patterns in some parts of the landscape historically.

Native people in the region today are working to revitalize traditional practices, and continue to gather a wide range of plants and other natural resources (A. Saubel pers. comm.). A summary of major cultural uses for a subset of native plants is included in the plant palettes in Appendix B.

Small patches of alkali meadow, marked by high surface concentrations of salts, occurred in areas with high groundwater levels and poorly drained, clay-rich soils (Nelson et al. 1917). Geologist William Blake, for instance, noted that "in many places along the river bottom, especially on the low ground, the soil is highly charged with salts, which effloresce on the surface, and form white crusts, preventing the growth of useful grasses" (Blake 1857). These seasonal wetlands were characterized by salt-tolerant plants such as saltgrass (*Distichlis spicata*), yerba mansa (*Anemopsis californica*), salt marsh fleabane (*Pluchea odorata*), and horned sea blite (*Suaeda calceoliformis*) (data from CCH).

Perennial wetlands were not common along this portion of the river, though a single large freshwater emergent marsh existed adjacent to the Santa Ana River in the Spring Brook drainage, near present-day Lake Evans (Goldworthy and Higbie 1871, Hall 1888b, Sanborn 1904, Brown and Boyd 1922). The marsh was dominated by tules (e.g., Olney's bulrush [*Schoenoplectus americanus*]), cattails (e.g., broadleaf cattail [*Typha latifolia*]), and rushes (*Juncus* spp.) and provided important habitat for species such as western pond turtle, common yellowthroat (*Geothlypis trichas*), and marsh wren (*Cistothorus palustris*) (Hall 1895, McLain 1899, Reed 1916, data from WFVZ).

**Figure 2.4. This ca. 1886 photo shows alluvial scrub within the Tequesquite Arroyo wash.** (Courtesy of the Museum of Riverside, Riverside, California)

## Upland habitats

The vast majority of the study area outside of the river corridor was dominated by a mix of **Riversidean sage scrub** and **forbland** (Fig. 2.5). Trees were virtually absent from the region outside of the riparian forest along the Santa Ana River (Nordhoff 1873, Minto 1878, Font and Brown 2011).

Sage scrub was dominated by California sagebrush (*Artemisia californica*), brittlebush (*Encelia farinosa*), buckwheat (*Eriogonum* spp.), white sage (*Salvia apiana*), and black sage (*S. mellifera*), and supported a diversity of other species such as desert wishbone-bush (*Mirabilis laevis*), purple threeawn (*Aristida purpurea*), chaparral bush-beardtongue (*Keckiella antirrhinoides*), bluewitch nightshade (*Solanum umbelliferum*), chia (*Salvia columbariae*), and lanceleaf liveforever (*Dudleya lanceolata*) (Hancock 1853a, Kelly et al. 2005, data from CCH). The pre-colonization composition of the forblands is not well documented, though evidence suggests these communities were characterized by a mix of perennial and annual forbs such as California poppy (*Eschscholzia californica*), Kellogg's tarweed (*Deinandra kelloggii*), Indian paintbrush (*Castilleja exserta*), lupine (*Lupinus* spp.), dwarf checkerbloom (*Sidalcea malviflora*), Johnny-jump-up (*Viola pedunculata*), and coastal tidytips (*Layia platyglossa*), along with grass species such as foothill needlegrass (*Nasella lepida*) and prairie junegrass (*Koeleria macrantha*) (Hall 1905, Reed 1916, Minnich 2008, data from CCH).

The extensive scrublands and forblands were described in some of the earliest written accounts of the region. Upon fording the Santa Ana River in January 1776, for instance, Friar Pedro Font described the hills to the west (possibly referring to the Pedley Hills) as "full of good grass both dry and green" (Font and Brown 2011). Traveling through San Bernardino Valley from Cajon Creek to Rancho Cucamonga in 1853, Blake (1857) observed "numerous herds of horses and cattle grazing upon the immense sheet of tall and luxuriant grass... variegated with an abundance of bright flowers" on the plains to the south. General Land Office (GLO) surveyors documented "prairie vegetation" and "wild sage" throughout the area in the mid-19th century (Hancock 1853a).

Evidence suggests that shrub vegetation on hillslopes was relatively dense and abundant, while shrub cover in lowland areas was relatively sparse. Describing the widespread Placentia soil types, for instance, Nelson et al. (1917) noted that the "higher areas" that are "rolling, or sloping... support a moderate to dense growth of... brush." The Jurupa Hills were "covered with brush from one end to the other" except for several "rather extensive grassy stretches" (Swarth 1908a). In contrast, GLO surveyors described the lowland areas in the southern and eastern sides of the study area as a "naked plain" with "sparse" vegetation (Hancock 1853b).

Riversidean sage scrub and forbland provided habitat for a wide range of species including mammals such as desert cottontail (*Sylvilagus audubonii*), black-tailed jack rabbit (*Lepus californicus*), and coyote (*Canis latrans*); birds such as California quail (*Callipepla californica*), California towhee (*Melozone crissalis*), Bell's sparrow (*Artemisiospiza belli*), cactus wren (*Campylorhynchus brunneicapillus*), Costa's hummingbird (*Calypte costae*), burrowing owl (*Athene cunicularia*), and Western meadowlark (*Sturnella neglecta*); reptiles such as common side-blotched lizard (*Uta stansburiana*); and numerous butterflies and other invertebrates (Gunn 1885; Twogood 1897; Swarth 1908a, 1908b; data from WFVZ and Vertnet).

While Riversidean sage scrub and forbland dominated upland areas, several other habitat types occupied small portions of the study area. A patch of **chaparral** dominated by chamise (*Adenostoma fasciculatum*) occupied the highest elevation areas of the Jurupa Hills, as well as other higher-elevation portions of the Jurupa Hills and adjacent mountain ranges outside of the study area (Kelly et al. 2005). A small **vernal pool complex** extended into the northern portion of the study area, contiguous with a somewhat larger vernal pool complex to the northwest of the study area (Nelson et al. 1915, 1917; Fairchild Aerial Surveys 1931). A number of small ephemeral or intermittent tributaries drained towards the Santa Ana River from the Pedley Hills, Jurupa Hills, and Box Springs Mountains, and likely supported narrow corridors of alluvial scrub (not mapped). Oak woodlands were not documented within the study area historically, though an isolated population of Palmer's oak (*Quercus palmeri*) located in the Jurupa Hills just north of the study area consists of a single clonal individual more than 13,000 years old (May et al. 2009).

> "The prairie is odorous with white sage and lupine in purple racemes, and the golden chalices of California poppies dash the dry, coarse foliage."
> — Magness 1899

**KEY MESSAGE**

The vast majority of the study area outside of the river corridor was historically dominated by a mix of Riversidean sage scrub and forbland.

**Figure 2.5. Riversidean sage scrub on the rocky slopes of Mt. Rubidoux, 1922.** (FL207521, Courtesy of the California History Room, California State Library, Sacramento, California)

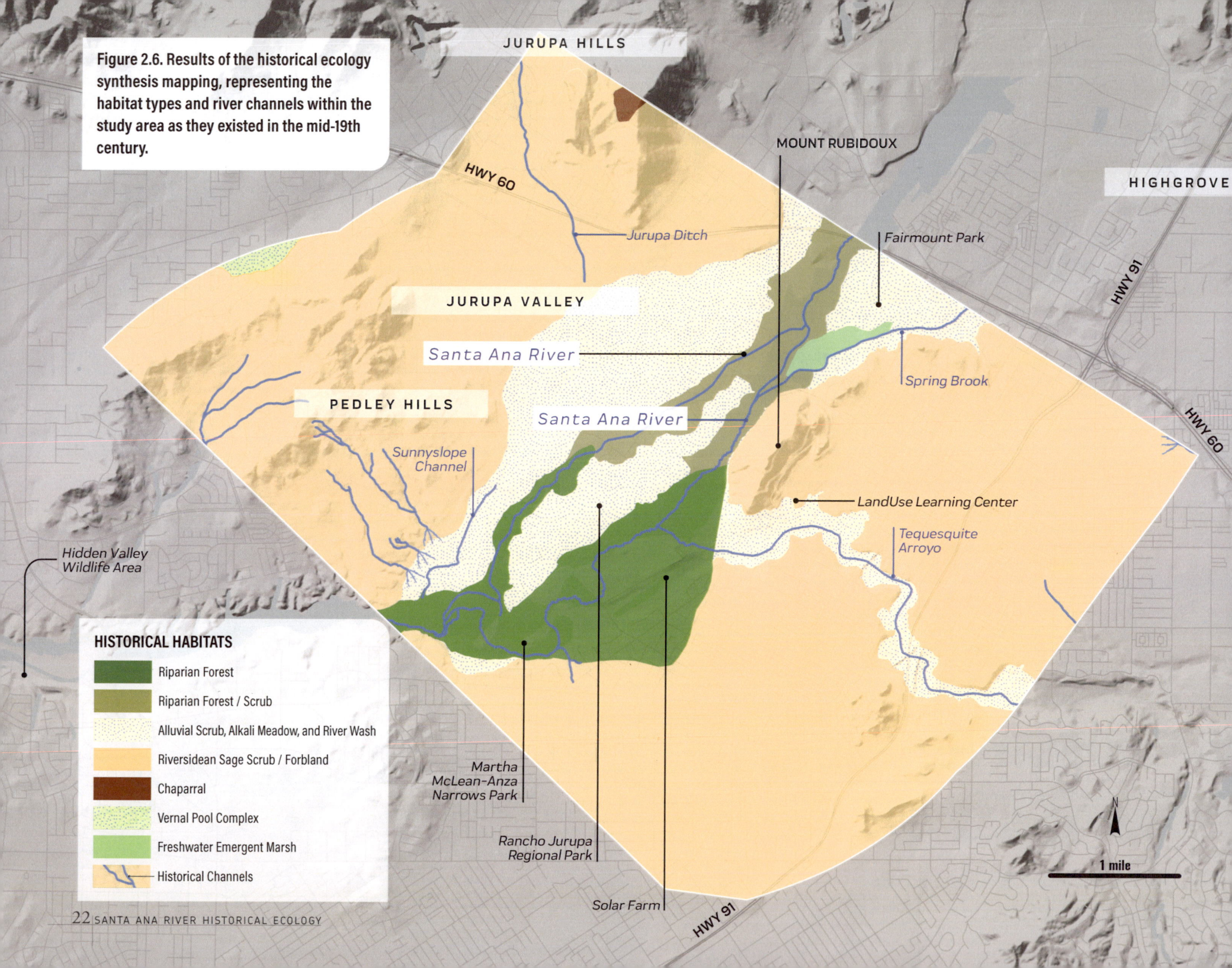

Figure 2.6. Results of the historical ecology synthesis mapping, representing the habitat types and river channels within the study area as they existed in the mid-19th century.

**Figure 2.7. Historical maps of the region.** Closewise from top left: mid-19th century map of Rancho Jurupa (USDC 1854-58), 1901 topographic map (USGS 1901), 1886 map of East Riverside Land Co. property (Dunlap 1886), 1888 irrigation map (Hall 1888b), 1915 soil map (Nelson et al. 1915).

# 3
# LANDSCAPE CHANGE and CURRENT CONDITIONS

**Irrigated beets, 1920.** (Courtesy of National Archives)

**Riverside and Santa Ana River.** (Imagery courtesy of Google Earth)

Major landscape changes began with the arrival of Europeans in the region in the 1770s and the resulting disruption of Gabrielino, Luiseño, Cahuilla, and Serrano cultures. During the late 18th and early 19th centuries, large numbers of Native people were enslaved at Spanish missions or asistencias, or later compelled to work on Mexican ranchos (Heizer et al. 1978). The Spanish intensified their settlement of the area beginning in 1818, bringing cattle and agriculture (Robinson 1957), as well as invasive grasses. Jurupa Rancho, encompassing much of the study area, was granted to Juan Bandini in 1838; a portion of this grant was later sold to Louis Rubidoux and became known as the Rubidoux Rancho (Greves 1876).

A massive flood in 1862, followed by several years of severe drought, decimated the livestock industry (Hall 1888a). With the decline of cattle ranching, ranchers began subdividing their land and turning to agricultural land uses. The Washington navel orange was introduced to Riverside in the early 1870s, and citrus cultivation quickly became the dominant industry in the region (Lech 2004). While the earliest water diversions in the region date back to Mexican settlements from the 1840s, construction of major irrigation canals began in the 1870s with the formation of the Southern California Colony Association and Riverside Land and Irrigating Company and the founding of the City of Riverside (Hall 1888a, Lech 2004). Urban and suburban development accelerated in the early 20th century, and particularly during the post-war era (Patterson 1996).

In addition to major changes in land cover type, invasive species introductions, nutrient additions, flow alteration and river channelization, and changes to climate and fire patterns have all altered vegetation communities. Invasive grasses have largely replaced native grasses and forbs. Major water projects to provide irrigation, flood control, and urban water resulted in a network of canals and a series of dams along the Santa Ana River, changing flood patterns and altering river habitats. The following sections briefly summarize the impact of some of these changes.

# Land cover change

**KEY MESSAGE**

Within the study area, 10,500 ac (4,250 ha) of native habitat (81% of the total area) has been converted to developed and disturbed land since the mid-1800s. Approximately half of the historical extent of riparian forest and riparian scrub has persisted.

We compared the historical ecology mapping (Fig. 3.2) to modern vegetation mapping (Fig. 3.3; Aerial Information Systems, Inc. 2012) to quantitatively assess landscape change over time. Table 3.1 shows the crosswalk between historical habitat classes and contemporary vegetation classification systems.

Within the study area, 10,500 ac (4,250 ha) of native habitat (81% of the total area) was converted to developed and disturbed land (Fig. 3.1, Table 3.2). Riversidean sage scrub and forbland is still the most abundant habitat type locally (1,250 ac [505 ha]), but 88% of this cover type has been converted to developed and agricultural lands. Most of the remnant sage scrub and forbland habitat is located in steeper upland areas such as the Jurupa Hills, Pedley Hills, and Mt. Rubidoux. The composition of these habitats, and in particular forbland/grassland, has changed considerably. Many grasslands in the region are now dominated by non-native grasses and forbs (Western Riverside County Regional Conservation Authority. 2003), though the relative extent of native and non-native grassland/forbland within the study area is unknown.

Within the river corridor, approximately half of the historical footprint of riparian forest and riparian scrub has persisted to today. However, other sites outside of this historical extent have been converted to riparian forest and riparian scrub, resulting in a total loss of only 32%. These vegetation types are combined in a single class in this analysis, as they are grouped in the modern vegetation mapping (Aerial Information Systems, Inc. 2012).

Riparian forest/scrub occupies most of the ~940 ft (285 m) wide corridor along the leveed portion of the Santa Ana River in the northern part of the study area. As in the historical landscape, the widest portion of the present-day riparian corridor is located in the southern portion of the study area, downstream of the leveed portion of the river. Contemporary riparian vegetation communities are composed of a mix of native species such as willow, cottonwood, and sycamore along with introduced species such as tamarisk (*Tamarix ramosissima*), arundo (*Arundo donax*), and castor bean (*Ricinus communis*).

Several of the habitat types historically found within the study area are either no longer present or nearly eliminated. The extent of freshwater emergent wetland has been reduced by 87%, from 53 ac to 7 ac, with the large historical freshwater marsh at the site of present-day Lake Evans being the most notable loss. Chaparral and vernal pool complex communities, which were uncommon in the study area historically, have disappeared entirely. Alluvial scrub, alkali meadow, and river wash were also not present in the modern vegetation mapping, although some smaller remnant or restored patches may exist within pockets of the riparian corridor. In addition, the Riverside-Corona Resource Conservation District recently coordinated restoration of alkali meadow habitat along Tequesquite Arroyo at Ryan Bonaminio Park.

Further analysis of the ecological quality and condition of the urban landscape is included in this chapter.

**Table 3.1. Crosswalk between historical habitat classes and contemporary vegetation classification systems.** Riparian forest and riparian forest/scrub were combined into a single historical class for the change analysis. Modern vegetation mapping was produced in 2012 for the Western Riverside County Multiple Species Habitat Conservation Plan (MSHCP) update. MSHCP Classes represent habitat types developed for the MSHCP mapping effort ("MSHCP_DESC" field in the GIS layer). The NVCS Classes column includes the National Vegetation Classification System associations, alliances, groups, and macrogroups present within the study area corresponding to each historical habitat class.

| Historical Habitat Class | MSHCP Class(es) | NVCS Class(es) |
|---|---|---|
| **Riparian Forest or Riparian Forest/Scrub** | Riparian scrub, woodland, forest | *Baccharis salicifolia* Alliance; *Platanus racemosa* Alliance; *Populus fremontii - Salix (laevigata, lasiolepis, lucida* ssp. *lasiandra)* Association; *Salix gooddingii* Alliance; *Salix gooddingii - Salix lucida - Populus fremontii* Association; *Salix laevigata* Alliance; *Sambucus nigra* Alliance |
| **Freshwater Emergent Marsh** | Meadows and marshes | Arid West freshwater emergent marsh Group |
| **Alluvial Scrub, Alkali Meadow, and River Wash** | Riversidean alluvial fan sage scrub | N/A (none within study area) |
| | Playas and vernal pools | N/A (none within study area) |
| **Riversidean Sage Scrub/ Forbland** | Grassland | California annual and perennial grassland macrogroup |
| | Coastal sage scrub | *Encelia farinosa* Alliance; *Encelia farinosa - Artemisia californica* Association; *Eriogonum fasciculatum* Alliance; *Opuntia littoralis* Alliance |
| **Chaparral** | Chaparral | N/A (none within study area) |
| **Vernal Pool Complex** | Playas and vernal pools | N/A (none within study area) |

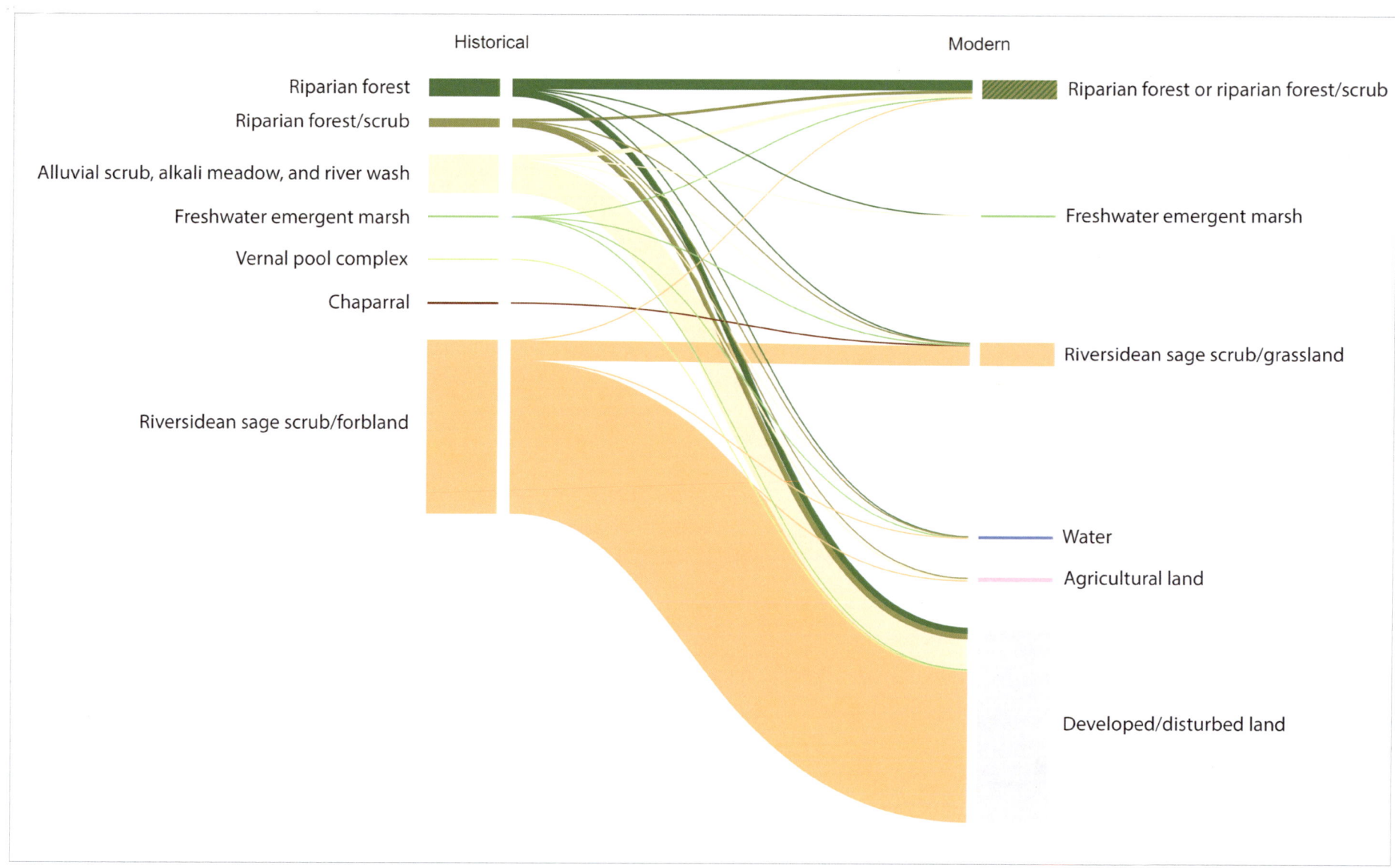

**Figure 3.1. Land cover and vegetation type conversion within the study area based on historical ecology results and 2012 vegetation mapping** (Aerial Information Systems, Inc. 2012). The bars on the left side represent the relative amount of each land cover type present within the study area in the mid-19th century, while the bars on the right represent modern land cover. Historical alluvial scrub, alkali meadow and river wash, vernal pool complex, and chaparral land cover are all no longer present within the study area in the modern mapping. The lines connecting the two sides of the chart illustrate the conversion "pathways" that have occurred over this period; the thickness of each line corresponds to the total number of acres that have undergone a given type of conversion.

Table 3.2. Overall change in the amount of land cover from the mid-19th century to 2012. The proportion of each historical habitat type that was converted to agricultural, developed, or disturbed land is shown on the right. These proportions will only total 100% when the entire habitat type has been converted to one of these highly modified land cover types, such as is the case for the vernal pool complex. In the case of other habitat types, some or all of their footprint was converted to another natural vegetation type, such as chaparral converting to Riversidean sage scrub. The riparian forest and riparian forest/scrub classes were mapped separately in the historically ecology mapping, but are aggregated in the modern vegetation type mapping.

| | Total acres historically | Total acres 2012 | Overall percent loss | Conversion to agricultural land | Conversion to developed/ disturbed land |
|---|---|---|---|---|---|
| Riparian Forest | 950 | 970 | 32% | – | 36% |
| Riparian Forest/Scrub | 475 | | | 1% | 60% |
| Alluvial Scrub, Alkali Meadow, and River Wash | 2,100 | 0 | 100% | 5% | 78% |
| Freshwater Emergent Marsh | 53 | 7 | 87% | – | 23% |
| Vernal Pool Complex | 37 | 0 | 100% | – | 100% |
| Chaparral | 19 | 0 | 100% | – | – |
| Riversidean Sage Scrub/Forbland | 9,360 | 1,250 | 87% | 1% | 87% |

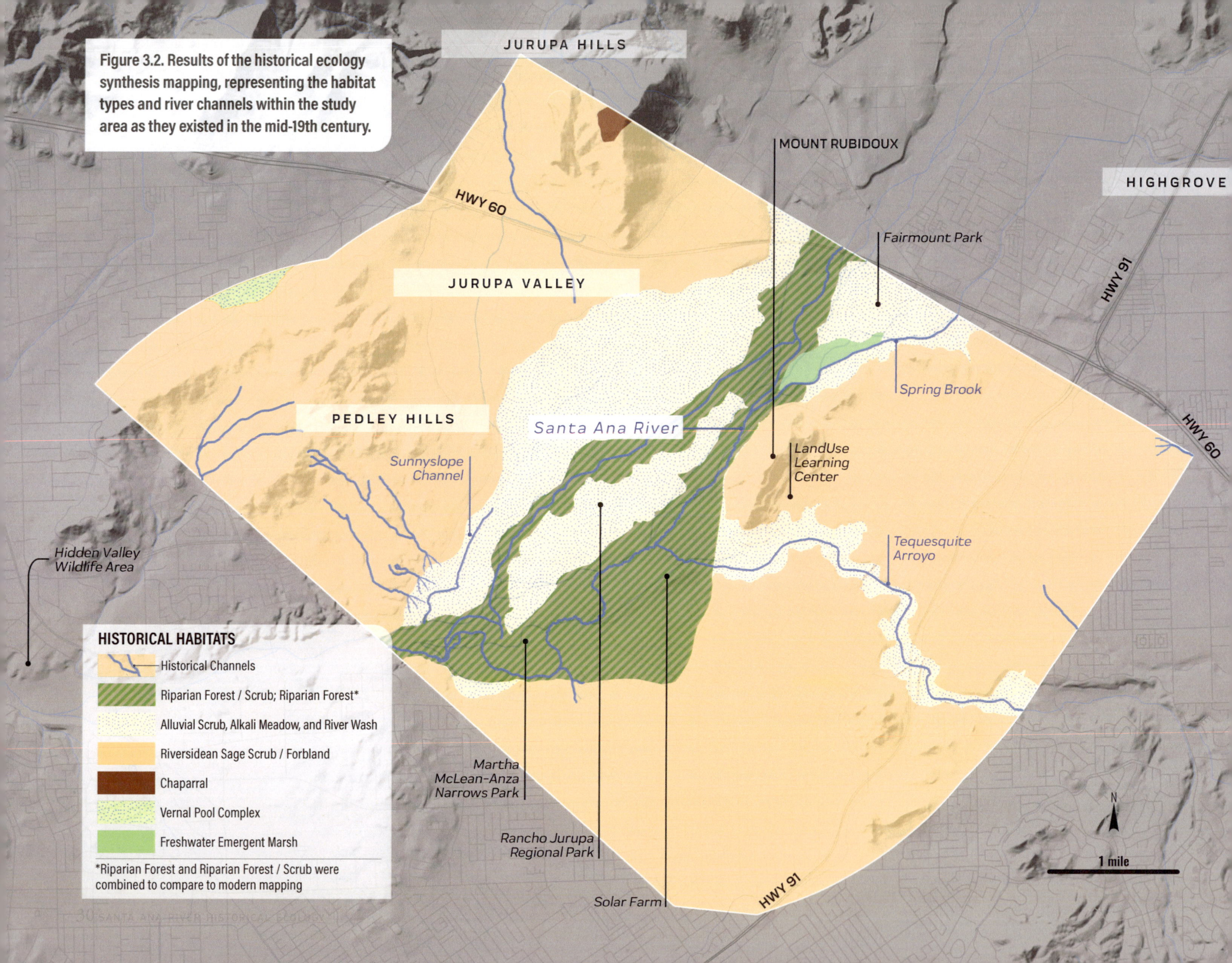

Figure 3.2. Results of the historical ecology synthesis mapping, representing the habitat types and river channels within the study area as they existed in the mid-19th century.

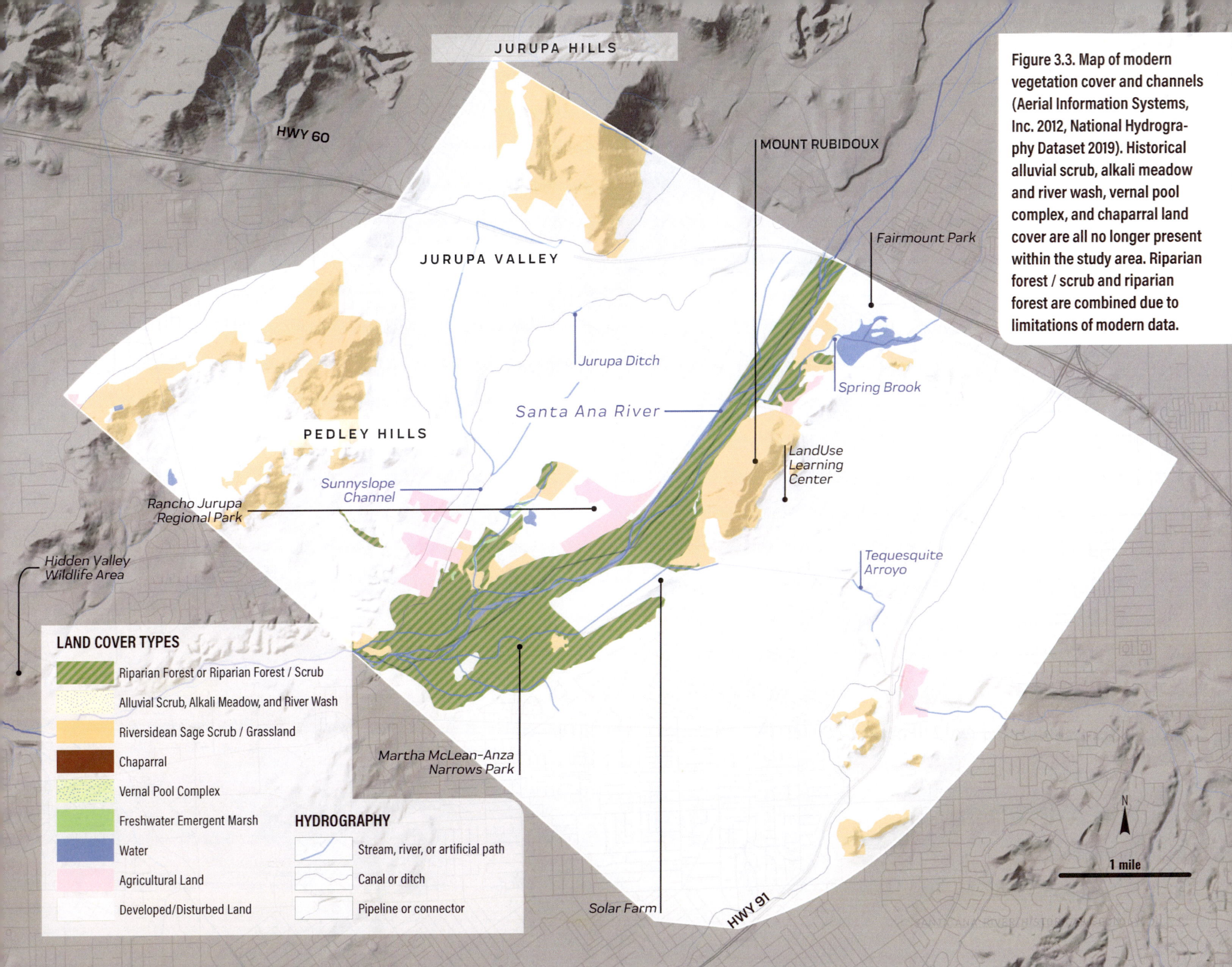

Figure 3.3. Map of modern vegetation cover and channels (Aerial Information Systems, Inc. 2012, National Hydrography Dataset 2019). Historical alluvial scrub, alkali meadow and river wash, vernal pool complex, and chaparral land cover are all no longer present within the study area. Riparian forest / scrub and riparian forest are combined due to limitations of modern data.

# Change to the Santa Ana River

Historically, the Santa Ana River within the study area had two core branches—an East Branch and a West Branch—and the relative size and flow of each branch varied through time. Before this region was settled in the late eighteenth century, the West Branch was likely the larger, more dominant path as water flowed down the Santa Ana River. However, evidence suggests that flood events in the late nineteenth century dramatically altered the course of the river, shifting the flow of water to the East Branch (Fig. 3.4).

Along the study area's short 4.5 mi (~7 km) stretch of the Santa Ana River, water flowed through nearly 9.5 mi (~15 km) of meandering channels. Today, the Santa Ana River flows through a single channel, roughly following the historical East Branch. This segment of the river has been partially straightened, channelized, and leveed, shortening the once sinuous river to less than half of its historical length and reducing channel complexity.

Much of what was historically the West Branch of the Santa Ana River is now developed residential land. The reach farthest downstream, where the West Branch once rejoined the East Branch, is now a small drainage surrounded by emergent marsh and riparian forest and scrub.

Upland from the river, water occasionally flowed through an additional 16 mi (26 km) of mapped waterways. Many of these arroyos spread through small distributary channels and were not directly connected with the Santa Ana River. Today, these smaller streams flow through a highly altered network of channels and canals, making up a total of almost 26 mi (42 km) of waterways. The West Riverside Canal, including its lateral extensions, the Riverside Canal, and the Jurupa Ditch, in particular, are major additions to the landscape that drastically modify the movement of water.

The Spring Brook Arroyo, which historically flowed through today's Fairmount Park and drained into a freshwater emergent wetland, has also been highly modified. Today, the channel is fed by recycled wastewater, and the creek is connected to the Santa Ana River through a culvert that runs under a levee.

While many streams flow through canals and water pipelines, some tributary arroyos still remain. The lower reaches of the Sunnyslope channel and several drainages in the hills to its north are still present, although partially channelized. The Tequesquite Arroyo also remains a major tributary to the south of the Santa Ana River. While its current channel generally follows its historical flow path, the Tequesquite Arroyo has been greatly modified, having been channelized or culverted along most of its reach.

In addition to these changes in channel form, river flows have also been altered, with impacts to instream condition and function. In California,

> "The Santa Ana River, in ordinary seasons, is dry for many miles below [Colton], where all the water is taken out to supply North and South Fork ditches. The waters of Warm Creek and other smaller tributaries, however, furnish a good stream again, which is taken out by the two Riverside canals to irrigate Riverside. In dry seasons these two canals take all the surface-water out of the river at these points, leaving the underflow to come to the surface below; but Spring Brook, which rises just northward of Riverside, replenishes the stream again."
>
> — Lindley and Widney 1888

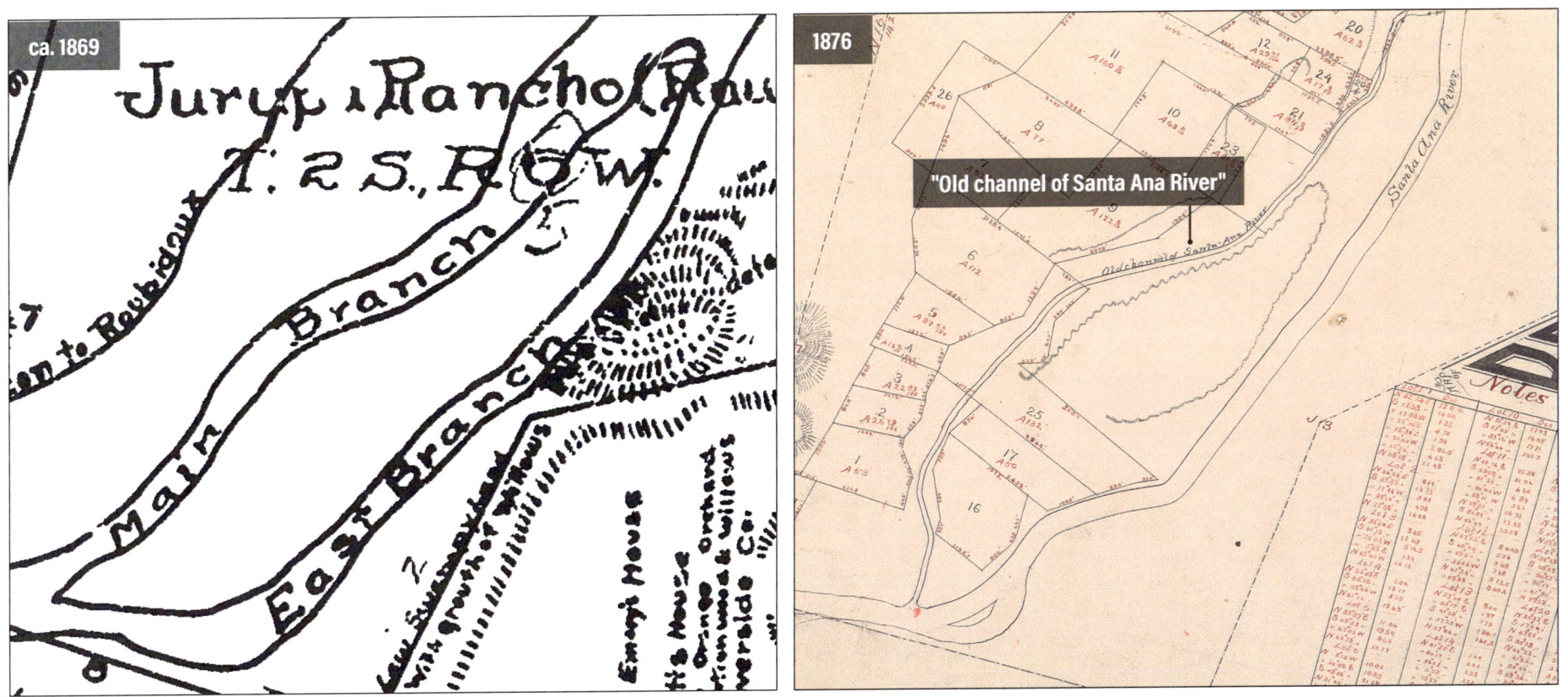

**Figure 3.4. The course of the Santa Ana River changed over time in response to flood events, as illustrated in these late 19th century maps.** The ca. 1869 map (left) labels the western branch of the river as the "main branch," while the 1876 map (right) labels this as the "old channel of Santa Ana River" and shows the eastern branch as dominant. (left: U.S. Surveyor General's Office 1878; right: Miller and Newman 1876).

key functional flows include peak flows, wet and dry season base flows, and recession flows linking the wet and dry season (Yarnell et al. 2020). Human water use and urbanization have altered each of these flow patterns, dating back to early irrigation diversions in the mid-19th century (see page 25).

The upstream Seven Oaks Dam (constructed between 1993 and 2000) stores water during high flow events for flood control and also diverts water for municipal use, resulting in a net reduction in flood flows (>4000 cubic feet per second) compared to historical conditions (San Bernardino Valley Municipal Water District and Western Municipal Water District 2004). Meanwhile, discharge from wastewater treatment plants together with urban and agricultural runoff has resulted in elevated base flow through the study area, although water recycling may begin to reverse this trend (SAWPA 2019). Together, these changes have resulted in fewer flood events and higher dry season baseflow (Fig. 3.5). Changes to peak flows can alter sediment transport, scour, and connection to the floodplain. Altered baseflow and wastewater discharge affect water quality, including temperature, nutrients, contaminants, and fine sediment, with impacts to instream habitat and aquatic species (SAWPA 2019).

Outside of the channel, lateral connection between the river and historical floodplain has been reduced by levees, which provide flood protection through a portion of the study area. Groundwater extraction has lowered groundwater levels outside of the river corridor, potentially limiting access to water for vegetation throughout the study area. Groundwater in Bunker Hill basin, which supplies 65% of water to Riverside, declined an average of eight feet per year between 1999 and 2018 (City of Riverside Public Utilities 2020).

Together, these changes reduce vertical, longitudinal, and lateral connectivity for the river, and may limit the ability of riparian tree species, such as cottonwoods, to establish. Chapter 4 discusses strategies to help manage these changes.

**Seven Oaks Dam, 2005.** (Courtesy CC 4.0, photograph by Steve Schumaker)

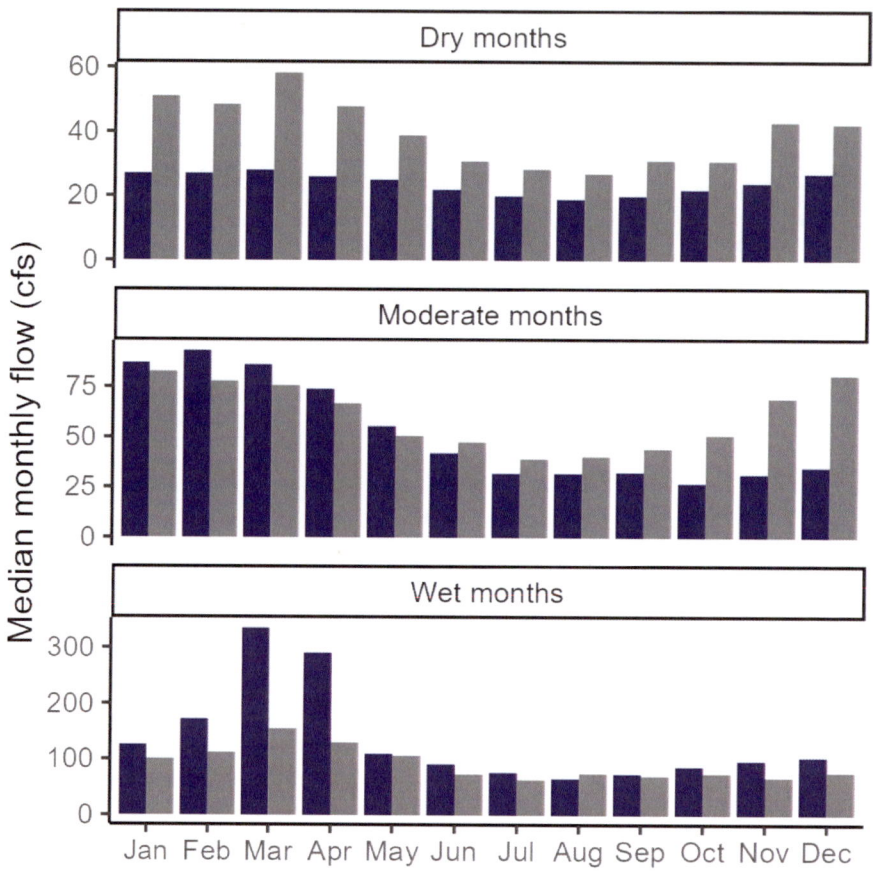

**Years**
- 1970-1990
- 2002-2023

**Figure 3.5. Dam construction and wastewater discharge have shifted flow patterns in the Santa Ana River.** This plot compares flows (cubic feet per second, cfs) over an equal period of record before and after the construction of the Seven Oaks Dam. Flows in dry months are elevated today compared to pre-1990 (top panel), and high winter flows are less common (bottom panel). (data from USGS gage 11066460: this analysis compares the driest 30% of each month prior to 1990 with the driest 30% of that month post 2002, and repeats for the middle 40% and the wettest 30%).

**Victoria Bridge spanning Tequesquite Arroyo, ca 1980.** (Courtesy Library of Congress)

### KEY MESSAGE

This segment of the Santa Ana River has been straightened, channelized, and leveed, shortening the river to less than half of its historical length and reducing channel complexity.

# Urban ecology assessment

While the Santa Ana River and its surroundings have changed drastically since the mid-19th century, this modified landscape still has the potential to support a healthy and resilient urban ecosystem. The Urban Biodiversity Framework (Spotswood et al. 2019) outlines the fundamental landscape elements that work together to support diverse, functioning urban ecosystems that have the capacity to both adapt to landscape changes and also generate various services and benefits for local communities, such as cleaner air and water, temperature moderation, increases in physical and mental health, and improved outdoor experiences.

The seven elements of urban biodiversity are **habitat patches, connections, matrix quality, habitat diversity, special resources, native vegetation,** and **management** (Fig. 3.6; Spotswood et al. 2019). We assessed current conditions, including key landscape features and needs, for supporting urban biodiversity using these seven landscape elements. The methods describing each analysis are detailed in Appendix A. The results from this assessment were then synthesized with the historical ecology findings to develop the recommended opportunities and strategies presented in Chapter 4.

**Santa Ana River adjacent to Riverside.** (Imagery courtesy of Google Earth)

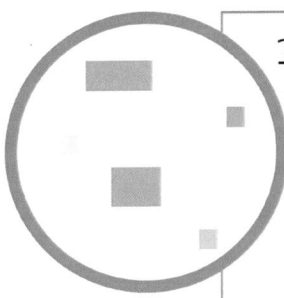

### 1 · PATCH SIZE

The size of a contiguous patch of greenspace in a city. We define patches as contiguous greenspaces of at least 2 acres in size.

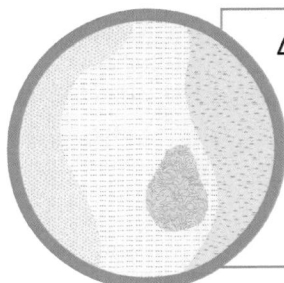

### 4 · HABITAT DIVERSITY

The type, number, and spatial distribution of habitat types within an urban area. Together, mosaics of habitats create diversity in habitat types at the landscape scale.

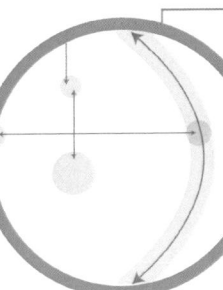

### 2 · CONNECTIONS

Features in the urban landscape that facilitate the movement of plants and animals. Connections include corridors (thin stretches of greenspace that promote linear movement) and stepping stones (sets of discrete but nearby patches that together promote connectivity across the landscape).

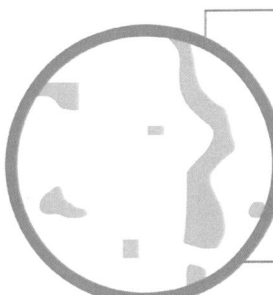

### 5 · NATIVE PLANT VEGETATION

Plant species long evolved in a specific geography (including nearby species that may be appropriate in the near future, given anticipated range shifts with climate change).

### 6 · SPECIAL RESOURCES

Unique habitat features necessary to support species' life history requirements, including large trees, wetlands, streams, and rivers.

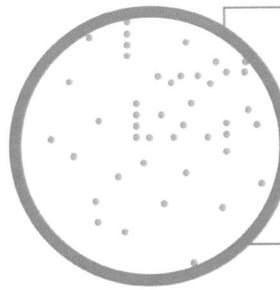

### 3 · MATRIX QUALITY

Habitat elements that support ecological process and movement in the urban matrix between patches of greenspace and corridors.

### 7 · MANAGEMENT

Human activities and planning that promote positive biodiversity outcomes.

**Figure 3.6. The seven elements of urban biodiversity work together to support resilient and diverse native species in urban areas.** These seven elements, drawing from the *Making Nature's City* urban biodiversity framework (Spotswood et al. 2019), guided the urban ecology assessment in this section. These elements are also described in more detail in subsequent pages and Appendix A.

### Habitat patches

Habitat patches are critical resources for biodiversity and a core pillar for fostering a healthy, functioning urban ecosystem. In this study, habitat patches were identified as continuous stretches of vegetation and greenspace greater than 2 ac (0.8 ha; see Appendix A for more details). These patches vary widely in their size and ecological quality, from city parks, sports fields, and private backyards to publicly protected and managed regional parks. While not all of these greenspaces are currently functioning as habitat, together, these open spaces represent opportunity areas that can be knitted together to create a cohesive network of core habitat to support a biodiverse urban ecosystem.

The cornerstone habitat patch within the study area is the Santa Ana River riparian corridor itself. This series of protected parks and open space forms an almost continuous greenspace complex, composed of wildlife reserves, parks, lands under conservation easements, outdoor recreation sites, sports fields, and other public facilities. Its large extent and connections up and down the watershed make it particularly valuable as an ecological resource, not only within the watershed, but across the region.

Outside of the riparian corridor, the largest protected patches of upland habitat are within Mt. Rubidoux Park and county parks across portions of the Jurupa Hills and Pedley Hills (Fig. 3.7). Smaller publicly and privately managed recreational parks within the study area, such as sports fields and golf courses, are other forms of open greenspace, with the potential to support wildlife through strategic planting and management regimes. This patchwork of smaller habitat patches between larger regional patches is important for maintaining biodiversity across the landscape.

Beyond the existing large, protected open spaces described above, several other potential patches remain unprotected and at risk of development or degradation, such as scrub habitat on Pachappa Hill and within large tracts of open space in the Pedley Hills and Jurupa Hills (Fig. 3.7). These undeveloped, largely intact tracts greater than 2 ac in size are the strongest candidates for protection and filling patch gaps.

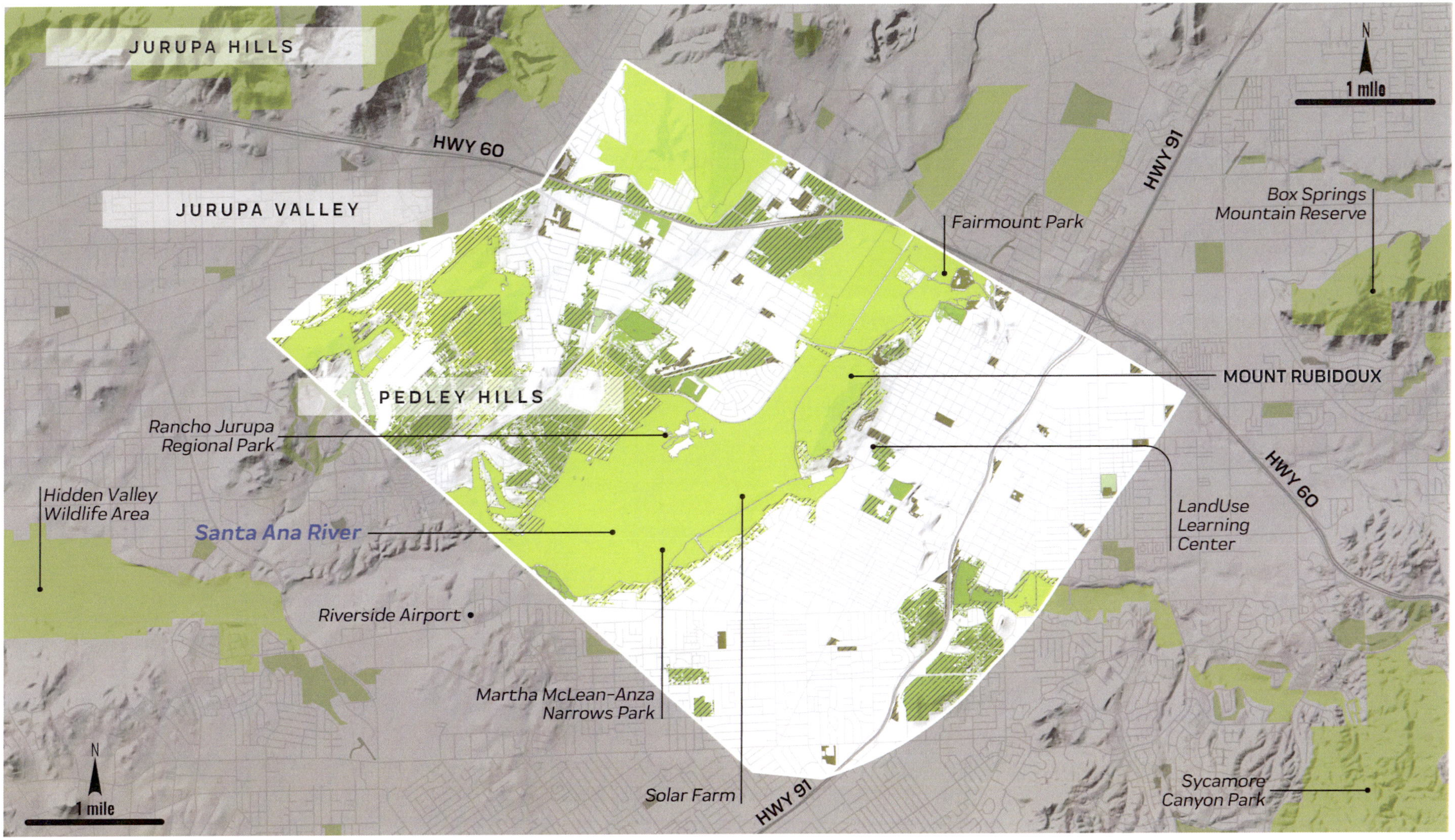

**Figure 3.7. Patches of protected and unprotected vegetated areas within the study area,** grouped into size classes that are ecologically significant: greater than 2 ac, 10 ac, and 130 ac, respectively (Spotswood et al. 2019). Patches were derived by identifying contiguous stretches of vegetation of at least 2 ac from 2016 land cover (EarthDefine 2016; see Appendix A for full methodology), and their protection status was related to protected areas databases (SCAG 2019, GreenInfo Network 2020).

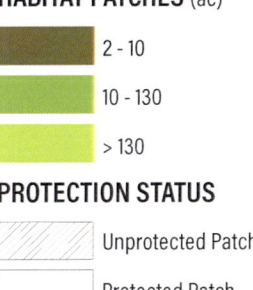

**HABITAT PATCHES** (ac)
- 2 – 10
- 10 – 130
- > 130

**PROTECTION STATUS**
- Unprotected Patch
- Protected Patch

## Connections

The Santa Ana River is a critical connectivity feature both within the study area and in the region. Surrounded by a series of greenspaces along much of its length, the river creates a continuous corridor from the mountains to the ocean. Within the study area, much of this corridor falls within protected land, creating an opportunity for coordinated restoration and management (see Fig. 1.3 on page 7).

When we analyzed connectivity across the study area, distinct patterns emerged on the landscape north and south of the Santa Ana River (Fig. 3.8). North of the river, particularly higher in the Jurupa Hills, there is a patchwork of intact habitat patches, agricultural land, and recreational greenspaces, which all serve as stepping stones for movement across the landscape. Additionally, a more extensive network of natural and unnatural channels and creeks north of the river also serve as important connecting features.

The most important barriers to ecological connectivity north of the river are located along major roads and highways where they cross between key ecological features. Important sites to restore connectivity are across State Route 60 where it meets the base of Pepe's Peak, Limonite Avenue to the north of the Jurupa Hills Country Club through which numerous tributary creeks flow, and Camino Real where it crosses between large patches in the Pedley Hills.

South of the Santa Ana River, the Tequesquite Arroyo serves as a key landscape connecting feature. However, there are two portions of the Tequesquite Arroyo that serve as particularly important barriers to landscape connectivity. The first major barrier is southeast of Ryan Bonaminio Park, and the other is where the arroyo crosses the Riverside Freeway (State Route 91).

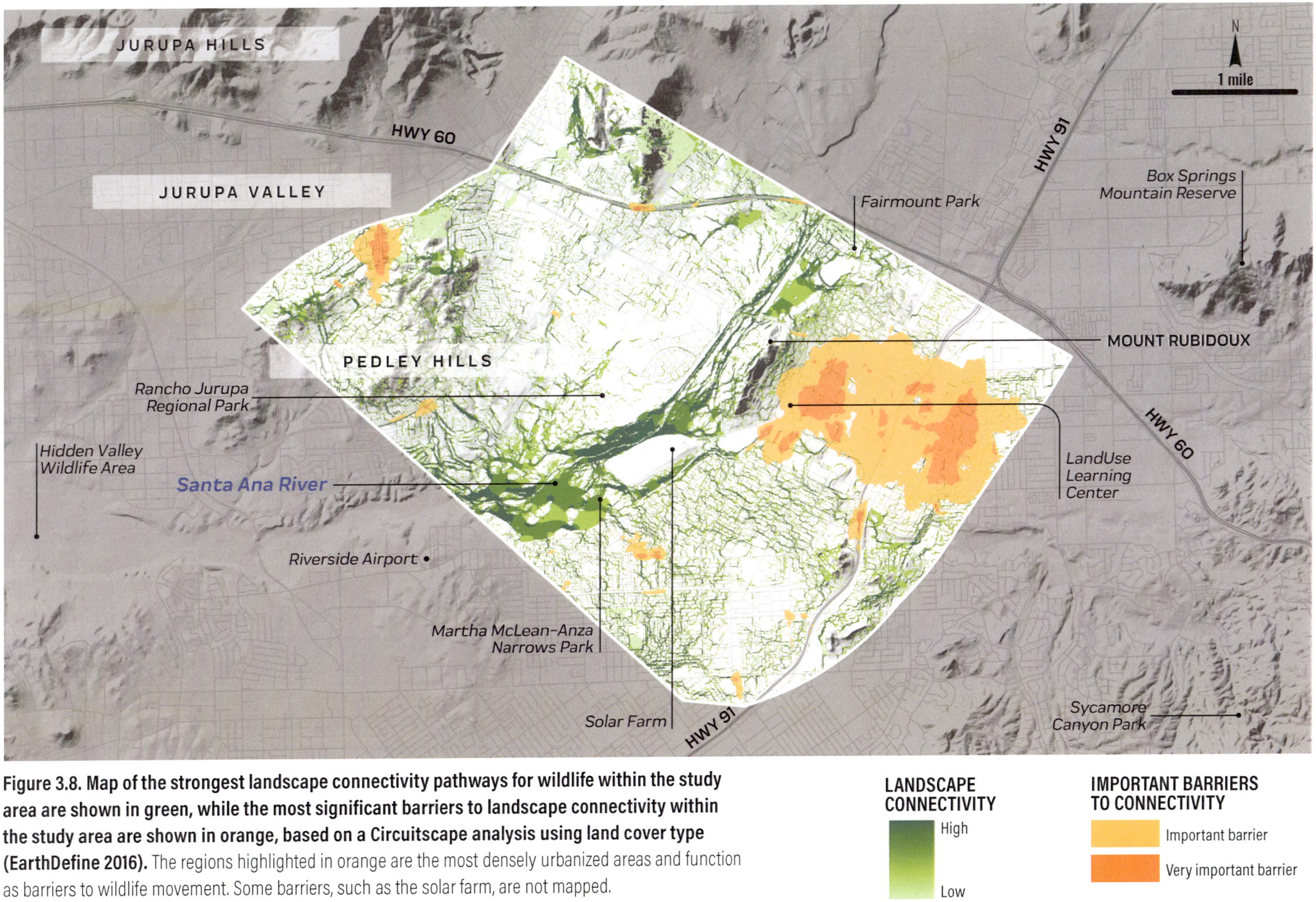

Figure 3.8. Map of the strongest landscape connectivity pathways for wildlife within the study area are shown in green, while the most significant barriers to landscape connectivity within the study area are shown in orange, based on a Circuitscape analysis using land cover type (EarthDefine 2016). The regions highlighted in orange are the most densely urbanized areas and function as barriers to wildlife movement. Some barriers, such as the solar farm, are not mapped.

## Matrix quality

Outside of larger habitat patches, vegetation is a vital landscape feature for supporting urban biodiversity. Focusing native urban greening around and between important habitat patches can increase the effective size of a patch or soften its edge, expanding the quality of resources, ecosystem functioning, and species movement beyond the boundary of the habitat patch. The quality of the urban matrix in turn amplifies the value of the habitat patches themselves, as patches surrounded by an ecologically rich and diverse urban matrix are able to support more species and higher numbers of individuals of those species by enhancing connectivity between patches (Malanson 2003, Baum et al. 2004, Evans et al. 2017).

Within the study area, shrub cover, historically the dominant cover type, is now relatively rare in the urban landscape. Figure 3.9 shows that while the upland protected areas, such as the Pedley Hills and Jurupa Hills, are still dominated by shrub cover, much of the urban matrix between these patches has low levels of shrub cover. In contrast, irrigation has allowed residential areas to support more tree cover than the historical landscape, as shown in green in Fig. 3.9. Even this tree canopy cover is unevenly distributed across the urban area, ranging from less than 10% to more than 30%.

While trees were historically rare in the upland landscape where the urban matrix now dominates, tree cover within the urban matrix plays an important role in cities today, generating numerous benefits for both urban residents and wildlife. Trees can provide habitat and food resources and enhance landscape connectivity for wildlife, and support ecosystem services and functioning (Matteson and Langellotto 2010, Bailey et al. 2019, Wood and Esaian 2020). Urban trees also provide protection from urban heat island (UHI) effect, among other physical and health benefits (Ziter et al. 2019). The Riverside-San Bernardino region is one of the most severely impacted regions in California for UHI. Urban-rural temperature differences range up to ~20°F (11°C) for census tracts within the study area and are likely to be exacerbated by climate change (Taha and Freed 2015).

In commercial and industrial areas of the study area, tree cover is relatively sparse, and asphalt, concrete, and other impervious surfaces dominate. These cover types provide little habitat value for native wildlife, and also exacerbate environmental threats to human wellbeing. Impervious surfaces lead to increased runoff during periods of high rainfall, leading to flashier flows and increasing local flood risk. Impervious surfaces also trap urban heat, augmenting urban heat islands that can exert significant impacts on human health (Heaviside, Macintyre, and Vardoulakis 2017).

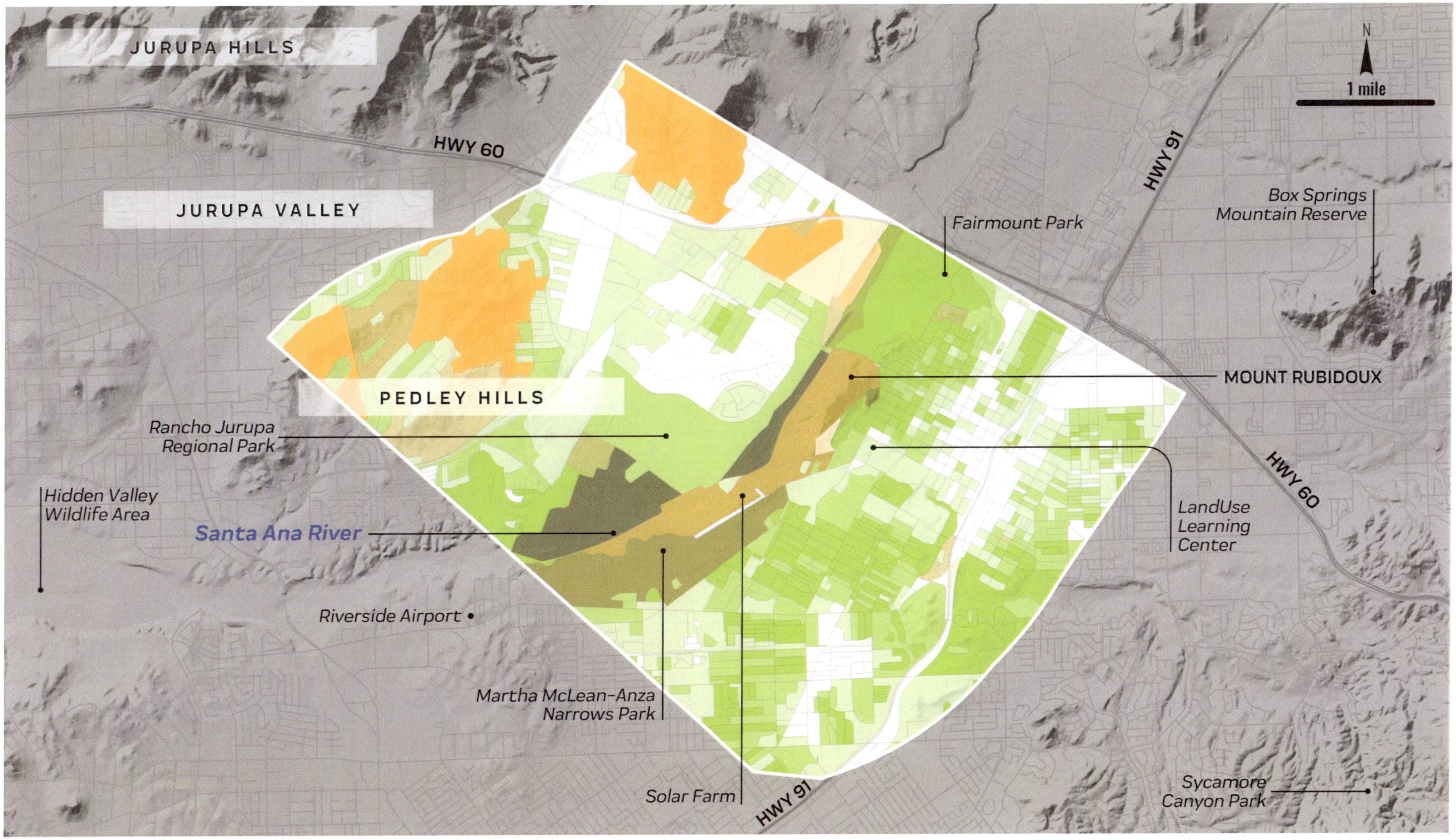

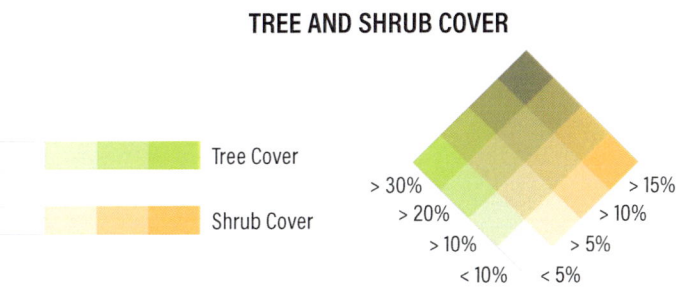

Figure 3.9. The percentage of total area within each U.S. census block composed of tree canopy cover (green) and shrub cover (orange). Census data are from the U.S. Census Bureau (2019) and land cover data are from EarthDefine (2016).

**TREE AND SHRUB COVER**

## *Habitat diversity and native vegetation*

The study area's uplands were historically dominated by Riversidean sage scrub and forbland, most of which has been converted to residential, commercial, and industrial land cover types. Remnant habitats remain largely on hilly terrain such as Mount Rubidoux, the Pedley Hills, and Box Spring Mountains, where development to date has been relatively limited (though large portions of these areas remain unprotected). In these intact habitat patches, an array of sage scrub vegetation alliances foster a diversity of habitats, supporting many species of plants and wildlife (Fig. 3.10). However, the historical scrub and forbland communities have also been impacted by non-native, invasive grasses, which are now dominant in the region's grasslands today.

A variety of riparian vegetation communities also exist along the Santa Ana River, including willow-cottonwood riparian forest, stands of California sycamore forest, and riparian scrub. Other historically significant habitats, including alluvial scrub, alkali meadow, and freshwater marsh, are now largely missing from the landscape. Like in the upland environments, invasive species also compete with native vegetation. Along waterways throughout the Santa Ana River watershed, arundo, castor bean, and tamarisk have replaced many native riparian forest species.

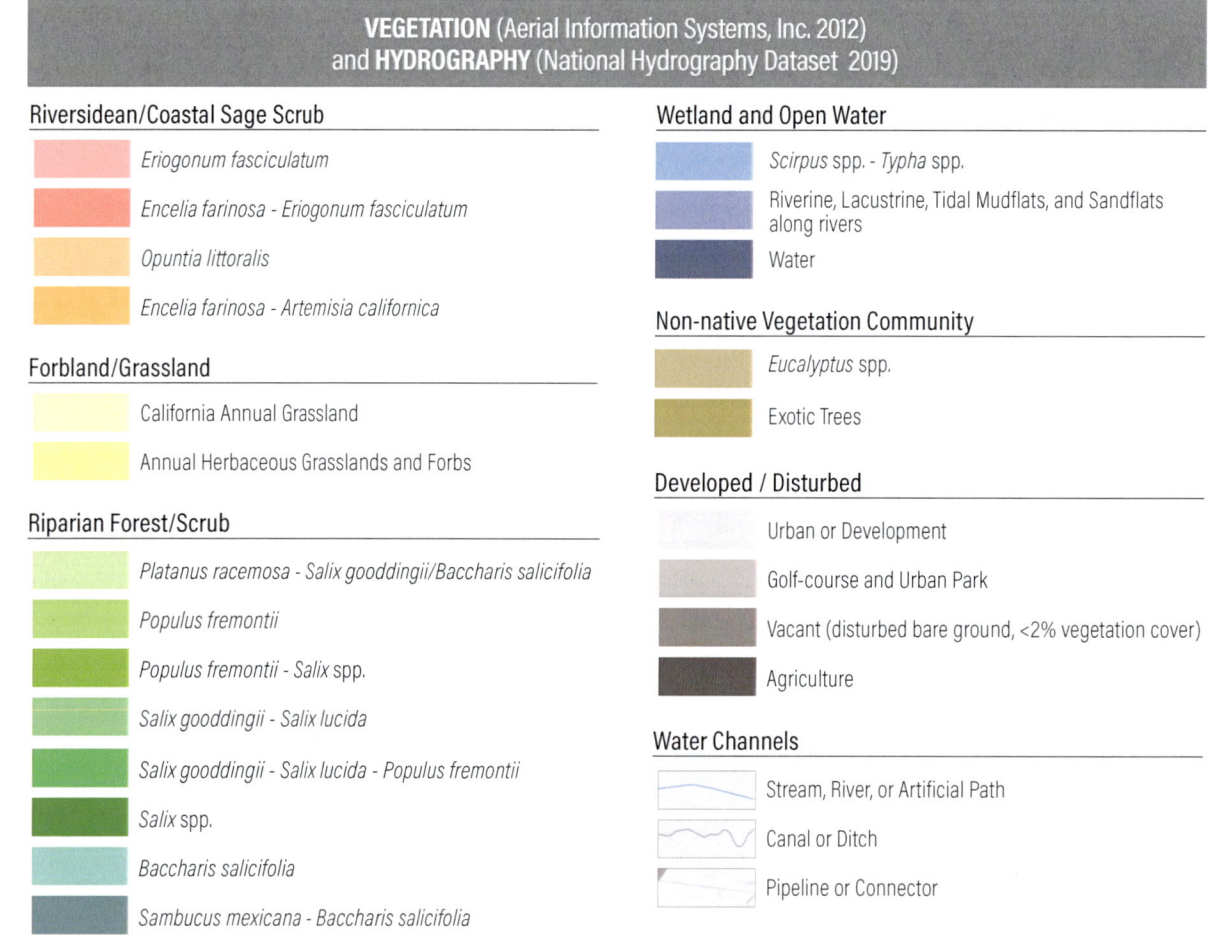

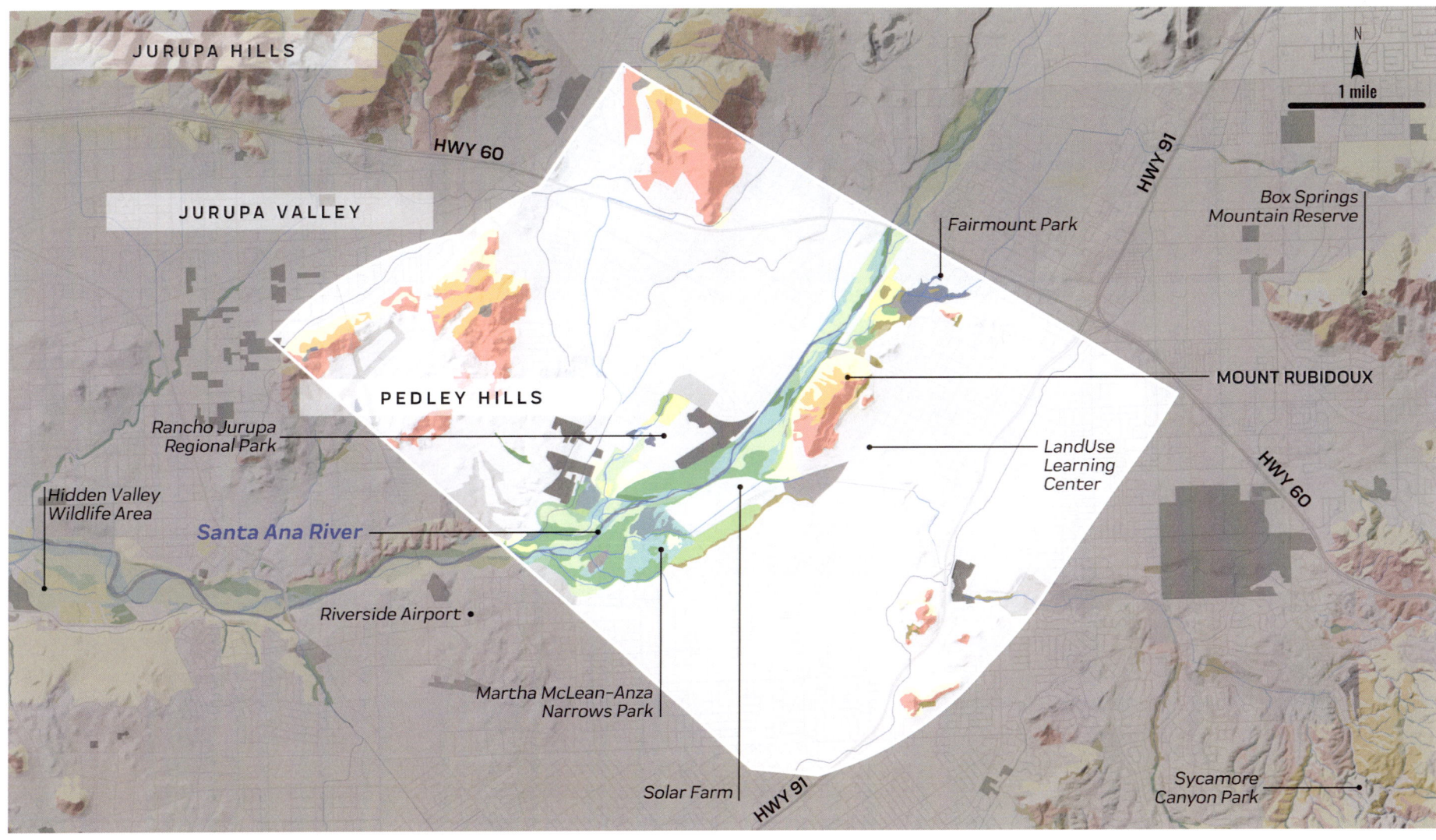

**Figure 3.10. A diversity of plant assemblages** are present in the study area, including many diverse communities of shrublands and riparian forest and scrub that were present historically. Because the historical ecology mapping could not be mapped to the detailed alliance-level classes shown in this map, many of these classes were grouped into the more general habitat type classes shown in Fig. 3.3 to enable comparison with the historical land cover. In addition to these natural habitat types, there are novel exotic tree communities now present in the study area. Much of the urban matrix vegetation is not mapped due to its high degree of impervious cover and variability in vegetation. (data from Aerial Information Systems, Inc. 2012, National Hydrography Dataset 2019)

## Special resources

Aquatic resources play a central role in the ecological health of the region, supporting a number of special status species such as the Santa Ana sucker and arroyo chub (*Gila orcuttii*). Riparian and wetland habitats that surround the Santa Ana River (Fig. 3.10) likewise support threatened and endangered species such as the least Bell's vireo (*Vireo bellii pusillus*) and southwestern willow flycatcher (*Empidonax trailii extimus*). Within the study area, southern cottonwood and willow riparian forest and southern willow scrub wetlands line the river and provide critical nesting habitat for many riparian wildlife species.

Outside of the river corridor, the remnant patches of scrub habitat and the open water in the location of historical freshwater marsh habitat near Camp Evans are also special ecological resources (Fig. 3.10). Remnant scrub provides habitat for a host of species, including desert cottontail (*Sylvilagus audubonii*), Costa's hummingbird (*Calypte costae*), Belding's orange-throated whiptail (*Aspidoscelis hyperythrus* ssp. *beldingi*), and many others. Due to the rarity of these habitat types in the present-day landscape and the distinct species assemblages they support, these areas make outsized contributions to the diversity and ecological function of the study area.

**Costa's hummingbird.** (Courtesy CC 4.0, photograph by Wendy Miller)

**Belding's Orange-throated whiptail.** (Courtesy CC 4.0, photograph by Wendy Miller)

## Management challenges & current efforts

Management actions can also impact the quality of habitat for biodiversity. The key management challenges affecting the study area include invasive species, nitrogen pollution, fire, soil compaction, pesticide use, climate change, and altered hydrology. Many of these land management challenges interact, exacerbating their impacts. As a result, considering potential interacting factors is often necessary to effectively manage ecological threats. Management challenges and actions typically influence ecosystems on large scales and are difficult to map, so in lieu of conducting a spatial analysis, we instead provide a few examples of the natural resource management challenges that overlap and interact within the study area, as well as current efforts in place to address them.

One complex local management challenge is fire. Fire frequency in the Santa Ana River watershed has risen dramatically in recent years, due to population growth and increased human ignitions, climate change, and exotic grass invasions (Sugihara 2006, Jin et al. 2015, SAWPA 2019). There have been at least 10 recorded wildland fires in the study area since 1950, ranging in size from 83 to 1,157 ac (34 to 468 ha), the majority of which can be attributed to human-ignition in dry months (CalFire 2017). The study area is subject to strong Santa Ana winds, which blow towards the coast during the fall when fuel is driest, creating some of the most severe fire conditions in the country (Sugihara 2006). Furthermore, areas of former Riversidean sage scrub have been invaded by non-native grasses, which thrive on frequent fires and contribute to higher fire frequency (Minnich and Dezzani 1998, Cione et al. 2002). To mitigate fire risk, the Santa Ana Watershed Project Authority and U.S. Forest Service have developed a Forest First Memorandum of Understanding (MOU) to support goals within the watershed including fuel reduction, restoration, and runoff control.

Fire risk in the area is compounded by the effects of invasive species and encampments of unhoused populations within the riverbed (GEI Consultants, Inc. and CWE 2020). Riparian woodlands have historically acted as barriers to fire spread due to high fuel moisture. However, invasion of arundo, an extremely aggressive and flammable non-native grass, has increased wildfire probability and intensity on the Santa Ana River in Riverside County (Bell 1998, Sugihara 2006). Meanwhile, encampments along the riverbed have contributed to increased rates of fire ignition. Fire management for the region therefore includes multiple strategies to address these interacting challenges. Efforts to eradicate invasive arundo have been largely successful in the northern portion of the study area above Mission Blvd (SAWPA 2019). Areas of removal priority include fire-prone, upland, and low-nutrient areas. Approaches to mitigate encampments' contributions to fire ignition and riverbed pollution include the City of Riverside's Wildlands Public Safety and Engagement Team (PSET) and Park and Neighborhood Specialists, adoption of City Ordinance 7606 to prohibit outdoor camping in wildland-urban interface areas, as well as Clean Camp Coalition trash services provided by the nonprofit Inland Empire Waterkeeper.

Other interacting management challenges include urbanization and soil compaction, which can increase runoff and limit opportunities for water to infiltrate, affecting both water supply and aquatic habitat. These impacts can be further exacerbated by more frequent extreme precipitation events caused by climate change. Efforts in the region to reduce impervious surfaces—such as implementing pervious pavement, bioswales, and urban greening—can help mitigate these effects. Meanwhile nitrogen pollution from car exhaust interacts with fire risk by encouraging the growth of invasive grasses in the region (Lu et al. 2003), which can then increase fire risk, particularly when combined with increased aridity under climate change. Statewide emissions reduction efforts currently in place will help to address overall nitrogen pollution levels, while sites facing non-native grass invasions can utilize mulch with a high carbon to nitrogen ratio (Allen et al. 1998; Cione, Padgett, and Allen 2002).

# CLIMATE CHANGE

Climate change projections for the Los Angeles region, including the Inland Empire, predict warming of 4-5°F (~2-3°C) by mid-century, as well as an increase in maximum temperatures, number of extremely hot days, droughts, and extreme weather events, including atmospheric rivers (Hall et al. 2018). Higher temperatures and more frequent drought may result in shifts in plant communities within the study area and will require human communities to adapt to extreme heat.

Although detailed species distribution modeling under climate change does not yet exist for the study area, regionally desert scrub vegetation types are projected to expand (Friggens et al. 2012). The species assemblages already present within Riverside may shift so that more xeric-tolerant species such as brittlebush, California buckwheat, and chaparral yucca (*Hesperoyucca whipplei*) become more dominant, particularly on south-facing slopes and other exposed areas. Sage scrub has some ability to adapt to variable precipitation, but additional stressors, including wildfire and invasive grasses, may limit the community's resilience (EcoAdapt 2016). To proactively manage for the changes in habitat suitability under climate change, managers should prioritize xeric-adapted plant assemblages in locations most vulnerable to extreme heat, and seek to identify locations that may remain cooler and could serve as refugia for less xeric-tolerant species.

Urban greening and strategic planting of street trees can help cities manage the effects of extreme heat by countering urban heat island effects. More details about strategic urban forest planting can be found in the opportunities and strategies section of this report.

**(left)** California hairstreak butterfly on buckwheat. (Courtesy CC 4.0, photograph by Virginia Rivers) **(right)** *Hesperoyucca whipplei*. (Courtesy CC 4.0, photograph by Tom Benson)

**Black sage, *Salvia mellifera*.** (Courtesy CC 4.0, photograph by Josh Jackson)

# 4
# OPPORTUNITIES and STRATEGIES

Illustrations by Vanessa Lee, SFEI.

# Introduction

Building on the historical ecology synthesis and the current conditions assessment, this section identifies opportunities and presents a series of recommendations to support improved ecological conditions within the study area. The recommendations are organized by habitat zone. **Habitat zones** are a spatial representation of target habitat types to help guide management and restoration. The habitat zones were developed by considering the study area's historical, present, and projected future conditions, including the city's historical ecology and native vegetation types, hydrology, soils, and climate.

Within each habitat zone, the level of human land use intensity varies, from urban development and high-impact recreational open space (high/moderate intensity) to naturalized, low-impact open space (low intensity). Specific actions should be tailored to the level of land use intensity required in a particular location.

Habitat zones are intended to be aspirational and provide a landscape-scale strategy for the types of habitats and ecosystem functions that should be considered at a particular site. For example, in most of the urban areas of Riverside, the historical land cover type was a combination of Riversidean sage scrub and forbland. While full habitat restoration is not a feasible goal in most urban areas, plantings in those areas could be designed to mimic Riversidean sage scrub and forbland, and could draw from that habitat zone's plant palette. Similarly, areas that were historically part of the riparian corridor along the Santa Ana River could be considered for restoration of cottonwoods, willows, and other riparian tree species, or might support alkali meadow or enhanced infiltration to groundwater.

This strategy envisions how the landscape can be planned at a high level, but local information should be evaluated before planning habitat restoration or enhancement projects at the site-scale. More details and specific recommendations for each of these habitat zones are provided below.

# General recommendations

Several of the recommendations span all of the habitat zones, focusing on the seven landscape elements that support urban biodiversity (Spotswood et al. 2019), described in Chapter 3. This suite of actions can be implemented across the urban landscape.

**Patches:** Plan buffers around existing habitat patches to create a gradual transition between sensitive habitats and more intense recreational land uses. These buffers can be composed of higher vegetation cover, and particularly native vegetation.

**Connections:** Improve habitat connectivity within and between parks. Consider creating "stepping stones" of habitat or focusing greening efforts between important natural patches where protection of a continuous corridor is not possible.

**Matrix quality:** Identify opportunities for strategic tree planting and urban greening that can support people and biodiversity. Evaluate whether soil compaction, urban heat, changed runoff, and infiltration may limit native species resilience and require additional mitigation strategies.

**Habitat diversity:** Design plant palettes and planting plans with the site's historical ecology in mind, such as by referring to the recommended plant palettes for each habitat zone below. Plant landscaped areas with both native plant diversity and vertical structural complexity, supporting a complex mix of groundcover, shrub cover, and tree canopy cover.

**Native vegetation:** Where possible, replace irrigated turf with native plantings, and select near-native street trees (i.e., trees native to the region). Manage invasive species in and around habitat patches that support high levels of native biodiversity.

**Special resources:** Prioritize habitat remnants identified in the habitat zones sections, particularly habitats within the riparian corridor and remnant wetlands.

**Management:** Incorporate wildlife-friendly design and management practices and engage the community in stewardship and decision-making. Examples of design and management practices include strategic fencing paired with dedicated wildlife crossings along roadways, reducing predation from domestic cats, minimizing use of pesticides and herbicides, use of bird-safe glass, and minimal and/or downward-facing outdoor lighting. Community participation in urban greening efforts and stewardship can expand the scale at which practices are adopted, amplify the value that the community derives, and promote the long-term resilience of the projects.

Connectivity is a particularly important landscape element to consider across the entire landscape, within and between each habitat zone. The Santa Ana River is the core connectivity feature both locally and regionally. While portions of the natural corridor along the river are intact within the study area, a large proportion of the area surrounding the river corridor is dominated by non-native plant assemblages.

Urban connectivity corridors that would link larger habitat patches are key places to focus greening and native landscaping efforts. The city can designate formal green corridors in concert with enhanced regional trail, pedestrian, and bicycle networks that link large recreational greenspaces. These green corridors, which can traverse existing natural features such as arroyos, can facilitate wildlife movement, linking upland habitats with each other and with the Santa Ana River corridor. Restoring the Tequesquite Arroyo is an especially critical opportunity to generate a major connectivity and recreational corridor, in addition to providing a multitude of other environmental services and benefits.

In the sections below, we outline more focused recommendations for each of the habitat zones shown in Fig. 4.1. See Appendix B for full plant palettes for each habitat zone.

**Figure 4.1. Map of recommended habitat zones.** Distinct levels of land use intensity are differentiated to help guide management decisions. Basemap from Esri world imagery (Esri et al. 2022).

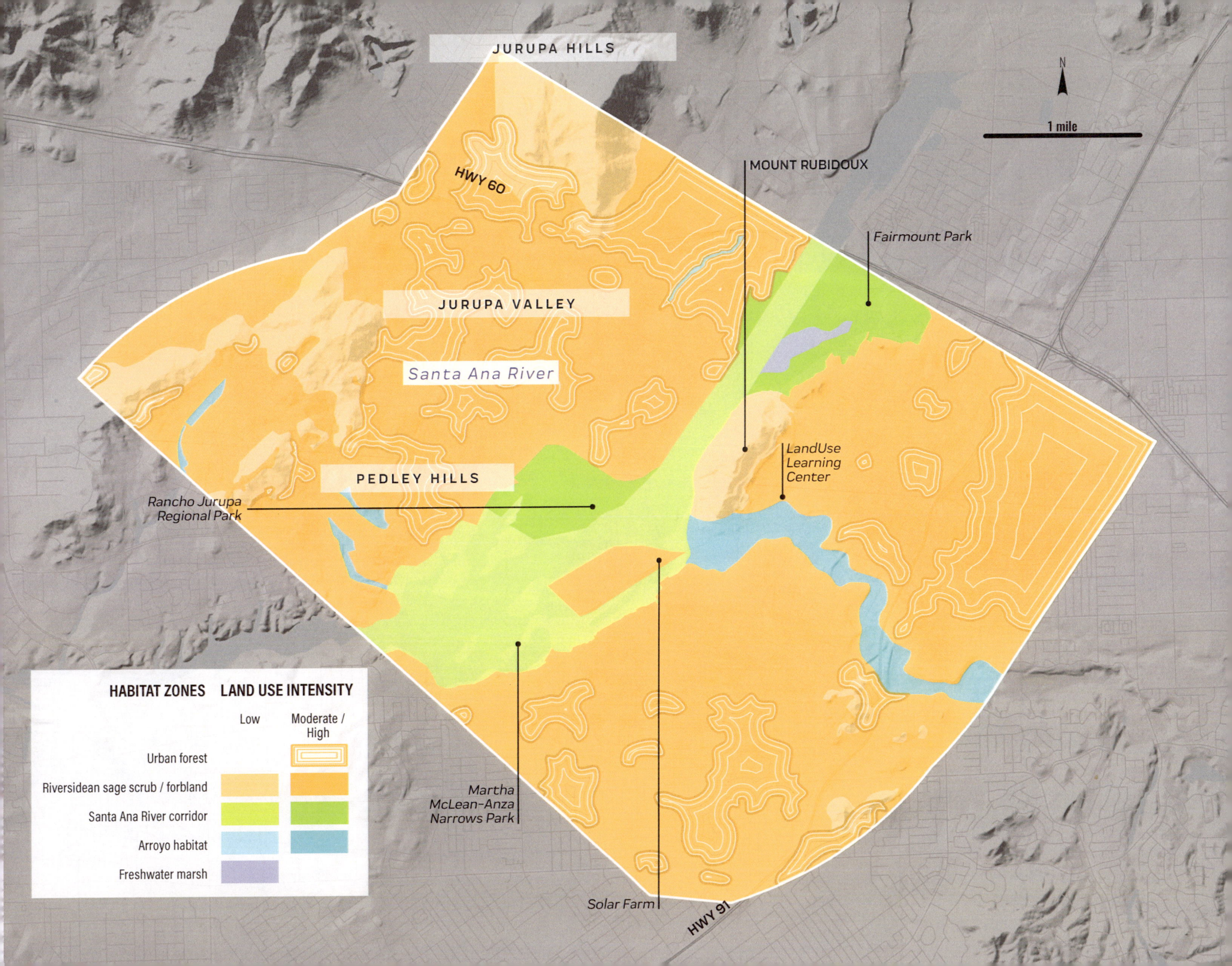

# Santa Ana River riparian corridor:
## Enhance the riparian zone as an ecological and recreational corridor

The Santa Ana River corridor was a dominant feature in the historical landscape and still offers opportunities for restoration of a unique and largely continuous corridor of habitat. The area mapped as "Santa Ana River corridor" in Fig. 4.1 includes portions of the historical extent of riparian forest, riparian scrub, alluvial scrub, alkali meadow, and river wash, and likely includes the primary opportunity areas for restoration of these habitat types.

As discussed in the landscape change assessment, the river corridor today is composed of a mix of riparian forest/scrub, as well as agricultural lands, recreational open space, and urban development.

Dams, levees, and urban land use have reduced connectivity between the river and its surroundings, and invasive arundo and other non-native plants have replaced or degraded a large portion of the native habitat in the riparian forest.

The riparian corridor has the potential to enhance connectivity across the region. With strategic management, it could support both public access and a variety of intact riparian habitats, including riparian forest, alluvial scrub, and river wash, as illustrated in Fig. 4.2. Recommendations for management focus on opportunities to restore a continuous corridor and the diversity of habitats present historically.

**Figure 4.2. This artistic depiction of a potential restored riparian corridor shows the patchy and varied vegetation characteristic of the historical riparian corridor.** Dominant overstory species could include Fremont cottonwood, willows (e.g., *Salix exigua, S. gooddingi, S. laevigata*), and California sycamore, while understory species could include mulefat and willow baccharis. (Illustration by Vanessa Lee, SFEI)

**Manage flows for ecological and geomorphic function and connectivity.** Maintenance of aquatic habitat can help protect the long-term viability of the Santa Ana sucker and other aquatic species.

- Upstream dams and urban water use have both resulted in altered flows. To support healthy ecological functioning in the Santa Ana River riparian corridor, management should seek to **manage for functional flows**, including high flows that can mobilize gravels and moderate flows to transport fine sediment.

- **Focused monitoring** can assess the impacts of elevated base flows and reduced high flows on local ecology. For example, monitoring efforts could assess whether gravels and cobbles are being adequately mobilized to maintain the condition of instream aquatic habitats.

**Re-envision riparian open space as a continuous corridor.** Creating a continuous riparian corridor would increase patch size and connectivity in the region, while also protecting the unique habitat and ecological functions provided by the river and riparian vegetation (Fig. 4.3).

- **Expand the width** of the riparian buffers to provide habitat for more species and generate diverse ecosystem services. As the Santa Ana River is a critical ecological and aquatic resource, the recommended edge buffer width is a minimum of **330 ft** (100 m) and ideally greater than **980 ft** (300 m) (Environmental Law Institute 2003). This riparian buffer should be a continuous belt surrounding aquatic resources, composed of permanent, and preferably native, vegetation. This buffer not only supports the riparian ecosystem's health, but also regulates the local temperature and microclimate, enhances water quality, and improves habitat quality for aquatic and terrestrial species alike.

- Management and planning should focus on **filling gaps** and **creating ecological gradients** in the buffer surrounding the sensitive riparian environment. While many open spaces along the river are wide enough to effectively buffer the riparian environment from other land uses, some sites adjacent to the river have low vegetation cover, degraded natural habitat, hard boundaries, or are used for activities that are disruptive to wildlife. For example, the Tequesquite Landfill Solar PV Project, sites within and surrounding Bubbling Well neighborhood, and Ryan Bonaminio Park can be priority areas for creating functional ecological gradients and reintroducing native landscaping.

- Where possible, **levee setbacks** and policies that allow for planting vegetation within levees can also support more ecological function and increase connection between the river and its surroundings.

**Restore native riparian vegetation and habitat diversity.** A focus on each of the distinct habitat types historically present in the riparian corridor can help support a variety of native species and ecological functions (Fig. 4.2).

- **Manage for varied habitats within the riparian corridor.** Historically, regular disturbance from the Santa Ana River would have maintained habitat diversity, including alluvial scrub, riparian scrub, and mature riparian forest. The patchy and varied habitat structure provides varied resources for wildlife.

- **Restore cottonwood and willow forests along the Santa Ana River,** particularly in areas with high groundwater, clay or silty soils, and low to moderate disturbance from flooding. Cottonwoods may need initial watering to establish, but should be planted in areas where they can readily reach groundwater (Fig. 4.4). Additional species include Fremont cottonwood, desert willow, Goodding's willow, red willow, California sycamore, mulefat, and willow baccharis.

- **Identify opportunities to restore alluvial scrub habitat** in areas that are periodically inundated by the Santa Ana River. Historically, alluvial scrub in the area was dominated by California scalebroom and was maintained by periodic scouring floods.

**Figure 4.3. The Santa Ana River sustains valuable riparian scrub and forest habitat along the Martha McLean Anza Narrows Park.** A series of protected habitat patches line the Santa Ana River and present an important opportunity for enhancement and creating a continuous connectivity corridor. (Photo by SFEI)

## Could this site be a cottonwood forest restoration site?

Was the site historically a cottonwood forest? → if YES → Is the site protected from annual scour? → if YES → Is the surface elevation within 6 feet of known groundwater? → if YES → Is watering possible within the first year to support establishment? → if YES → **YES, this site is a potential location for successful cottonwood forest restoration**

**Figure 4.4. Flowchart for identifying potential cottonwood forest restoration sites.** Final identification of appropriate sites will likely require site-specific data collection to assess groundwater level. Adapted from Stillwater Sciences (2007) and Laub et al. (2020).

- **In areas with high intensity land use** that are no longer directly connected to the Santa Ana River, look for opportunities for **selective reintroduction of native cottonwoods, willows, and other riparian vegetation.** Prioritize locations where these species will be supported by groundwater. When riparian plantings are not feasible, native landscaping can help soften the edges between the riparian habitat and human land uses. Rancho Jurupa Park, the Bubbling Well neighborhood, and undeveloped patches along 34th street and Highway 60 represent potential opportunities for expanding connectivity of the corridor through revegetation.

- **Conduct groundwater assessments** to inform the selection of planting sites within the urban matrix or open space adjacent to the river. Many riparian species are dependent on ready access to groundwater, historically complemented by periodic flooding. In high land use areas in particular, assessments of groundwater and soils should inform planting and site selection.

- **Manage habitat for a diversity of focal wildlife** in accordance with the Western Riverside County Multi-Species Habitat Conservation Plan, such as for the least Bell's vireo, southern willow flycatcher, yellow-breasted chat (*Icteria virens*), and yellow-billed cuckoo (*Coccyzus americanus*).

## Reduce human disturbance of the river.

- Creating **designated access points** in current high-use areas could help minimize the impact of the many unsanctioned river access routes (Fig. 4.5).

- These access points should be complemented with **well-marked, maintained trails** that clearly deter unauthorized uses and protect sensitive areas of the river as higher quality habitat. Where necessary, use wildlife-permeable fencing or barriers, such as post-and-rail fences or boulders.

**Figure 4.5. The Santa Ana River is commonly accessed through informal trail networks.** This fence along a trail at Martha McLean Anza Narrows Park, designed to keep visitors on designated trails, has been cut through to create unsanctioned access to the river. (Photo by SFEI)

- **Collaborate with the unhoused community** currently living in the river corridor to find supportive solutions and **legal camping sites** located away from sensitive ecological resources. Promoting sustainable housing solutions outside the river corridor will reduce fires and other human disturbances.

# Arroyos:
## Consider restoration and infiltration opportunities

Tributary arroyos connect the uplands with the Santa Ana River and can form natural corridors to help wildlife and plants move across the watershed. Healthy tributaries promote the overall health of the river and the riparian corridor; they are also important habitats in their own right, providing conditions that some species depend on entirely or at specific stages in their life cycles. Within the study area, tributary arroyos historically supported alluvial scrub and alkali meadow habitat.

Many of the tributaries in the study area are physically and ecologically imperiled today, following many years of modifications for flood control and urban development, as well as the accumulation of invasive species and debris. Tequesquite Arroyo, the largest arroyo within the study area, is largely underground or routed through narrow ditches through the City of Riverside. Sunnyslope Creek has been degraded due to an accumulation of debris, silt, and vegetation growth in its downstream section. Spring Brook Arroyo was highly altered by the development of Lake Evans, and connects to the Santa Ana River via a culvert through the levee. Other historical tributaries have likewise been channelized and culverted.

Recommendations focus on opportunities to restore key infiltration and habitat functions associated with the historical arroyos. Arroyos have the potential to be critical connecting features on the landscape, both for people and wildlife.

**Reintroduce arroyo function through infiltration.** Even where restoration opportunities are limited, enhancing infiltration can recapture some of the arroyos' historical functions.

- Identify potential locations to **restore infiltration** by naturalizing channels, removing hard channel beds, and daylighting arroyos that have been piped underground.

- Within high land use intensity areas, **green stormwater infrastructure,** such as bioswales and rain gardens, can be installed along the course of historical arroyos to help absorb and direct flow during high intensity rain events.

- **Restoring and greening the Tequesquite Arroyo** is a regional priority for flood protection, water quality, and groundwater recharge (Placeworks 2018). The historical course of the Tequesquite is underlain by coarse alluvial deposits that may still support infiltration. Much of its extent has been converted to open space, sports fields, and parking lots where daylighting, restoration, or green infrastructure are particularly feasible.

**Connect wildlife habitat through restoration of native species and habitat types.** Although human land uses, such as residential and commercial development, roads, and sports fields, have encroached on arroyos, traces of these features are still present in the vegetation makeup (Fig. 4.6). Arroyos represent natural pathways for wildlife to move across the landscape from the ridgelines to the Santa Ana River.

- Consider the appropriateness of **a distinct species assemblage** along historical arroyos, and seek opportunities to restore a vegetated corridor along the path of historical arroyos within existing open spaces or backyards (Fig. 4.7). Groundwater and soils may still support distinct species assemblages along these historical features, which can help create connections for wildlife. Areas with sufficient moisture may support riparian species such as rushes (e.g. *Juncus* spp.), sedges (e.g., *Carex* spp.), willows (*Salix* spp.) and mulefat. Large-scale restoration of alluvial scrub vegetation communities may be infeasible in areas that no longer experience periodic scouring floods, though scalebroom and other elements of alluvial scrub habitats could be supported along arroyos in areas where active maintenance activities are possible. See Appendix B for a list of plants appropriate for alluvial scrub or riparian forest communities, as conditions permit.

### Connect people to the Santa Ana River through its tributaries.

Urban streams can provide important opportunities to connect people with their surroundings and create a sense of place.

- Consider aligning low-impact recreational and active transportation corridors, such as trails, pedestrian and bike routes, and green streets, with arroyos and drainages. While human uses should be planned and designed to minimize impact to the watercourses, these tributaries can serve as gateways to the river itself, reconnecting urban residents with the river and its surrounding greenspaces. Furthermore, restoring arroyos alongside low-impact recreational and active transportation facilities can deliver greater benefits for local residents, such as improving stormwater infiltration and water quality, creating recreational opportunities that improve mental and physical health, and mitigating urban heat island effect.

- Invest in sustained management and monitoring to manage human impacts. Important management actions include the removal of exotic species, mitigation and remediation of unsanctioned trails, trash management, and sediment management.

**Figure 4.6. (top) The lower extent of Spring Brook Arroyo in Fairmount Park,** where a dense riparian community is still supported along the wash. (Photo by SFEI)

**Figure 4.7. (right) A small tributary (lower-right of photo) to Sunnyslope channel that has been straightened, channelized, and culverted under Riverview Drive.** This site presents an opportunity to restore a Santa Ana River tributary. These tributaries can act as vital connecting features on the landscape, provide habitat for wildlife, and increase filtration and infiltration of storm flows. (Photo by SFEI)

# Wetland habitats:
## Restore rare features with high conservation value

Within the study area there were historically patches of alkali meadow, a small amount of vernal pool habitat, and a freshwater emergent marsh near Camp Evans. Wetlands provide a distinct set of resources for wildlife and boost habitat complexity across the landscape. They also support plant species and invertebrates not present elsewhere. Wetland features have almost entirely disappeared from the landscape, so even small areas of wetland restoration can provide large benefits.

**Restore the historical freshwater emergent marsh** through the Camp Evans Master Plan effort. Historically the marsh likely reached from the Santa Ana River to the midpoint of Lake Evans, and potentially reached even further northeast up the Spring Brook Arroyo and into the area where Fairmount Golf Course is located.

- Opportunities to restore the freshwater marsh exist both **northeast of Lake Evans** where Spring Brook Arroyo feeds into the lake and **southwest of Lake Evans** where the lake drains back into Spring Brook Arroyo and flows through the site of the historical wetland (Fig. 4.8). Restoring wetlands at these sites can provide additional benefits, particularly during storm overflow events when these sites currently flood.

**Coordinated actions should be taken in and around Lake Evans to improve water quality and reduce pollution sources and control the invasive fishes currently residing in the lake.**

- Policies and protections include creating a buffer around the lake and banning the use of pesticides and chemical fertilizer in the surrounding landscape, especially on the nearby golf course.

**Identify opportunities to restore and protect alkali meadow, a locally rare and specialized habitat type.** Sites with relatively fine-grained soils and high groundwater levels may have potential for supporting alkali meadow habitat. Saltgrass was likely the dominant species in alkali meadows historically, though these wetlands supported a variety of salt-tolerant plants (see Appendix B).

- Work closely with local experts to identify suitable alkali meadow restoration sites, select appropriate suites of species, and set restoration targets. Historically, alkali meadow occurred in small patches in a number of areas within the Santa Ana River corridor and along tributaries like Tequesquite Arroyo.

- Alkali meadow habitat has been recently restored along the Tequesquite Arroyo at Ryan Bonaminio Park. Implementing long-term monitoring, including identifying and managing threats such as invasive species, can help maintain the success of this restoration project.

**Figure 4.8.** Spring Brook Arroyo flows along its historical watercourse, but drains today into Lake Evans (top), which has replaced a freshwater emergent marsh. Water in Lake Evans comes from a combination of Spring Brook Arroyo flows, recycled water, and pumped groundwater. The outlet channel below Lake Evans (bottom) supports riparian species, but its ecological and physical condition and functioning are degraded by the invasion of non-native species. Restoring portions of the freshwater marsh above and below Lake Evans and managing invasive species along Spring Brook Arroyo can greatly enhance the ecological value of the park. (Photos by SFEI)

# Riversidean sage scrub and forbland:
## Seek opportunities to integrate native species

The uplands of Riverside and Jurupa Valley historically hosted extensive areas of shrubland and forbland, dominated by sages (*Salvia* spp.), sagebrush, California buckwheat, brittlebush, and a wide variety of native grasses and wildflowers such as California poppy, dwarf checkerbloom, and small flowered needlegrass. Several native birds and butterflies rely specifically on these plants as sources of food and shelter.

While only small, confined tracts of this habitat remain today, its component species can be incorporated into the urban landscape. Buckwheat, sagebrush, and native sages (especially *Salvia mellifera* and *S. apiana*) are highly drought-tolerant, fast-growing shrubs that already grow successfully in urban gardens in the area. Riversidean sage scrub and forbland can be integrated into present-day open space (Fig. 4.9), as well as urban and managed parkland areas, such as golf courses and cemeteries. In areas that are more actively used by people (urban and parkland areas), native shrub cover can be a feature in ornamental plantings and road medians.

**Figure 4.9. An artistic representation of restored Riversidean sage scrub and forbland** intermixed with low-impact land uses, including recreational sports complexes, a bike path, and a hiking trail. Dominant scrub species include California sagebrush, black sage, California buckwheat, and California brittlebush. (Illustration by Vanessa Lee, SFEI)

**Protect remnant tracts of Riversidean sage scrub.**

- There are several tracts of remnant Riversidean sage scrub habitat that remain unprotected in the study area, including in the Pedley Hills, Jurupa Hills, and Pachappa Hill. **Protecting these undeveloped areas**, through either fee simple acquisition or conservation easement, can help maintain the existing scrub habitat within the study area.

**Implement strategic planting and management** regimes on other publicly and privately managed open spaces within the study area, such as parks, sports fields, golf courses, and cemeteries (Fig. 4.9).

- **Strategic plantings** include converting sports fields' and golf courses' out-of-play areas to native landscaping and planting native and near-native trees in and around parks, cemeteries, and sports fields where they can provide shade for visitors (see Fig. 4.10).

- **Promote wildlife-supporting management practices**, such as removing fencing or replacing it with wildlife-friendly fencing; reducing or eliminating use of fertilizers, pesticides, and irrigation; using wildlife-friendly lighting practices; and implementing integrated pest management and drought management plans.

**Prioritize planting native or near-native scrub and forbs** strategically throughout the city's urban matrix.

- **Prioritize native plantings near existing patches**. Historically dominant species include California sagebrush, brittlebush, California buckwheat, black sage, and California poppy. Planting native vegetation near habitat patches can buffer habitat from surrounding land use and improve habitat quality for wildlife.

- **Identify potential habitat corridors** that link patches and can also provide benefits for people. Planting native trees and shrubs along active transportation corridors, slow streets, and residential yards, as shown in Fig. 4.10, can provide urban cooling, as well as mental and physical health benefits.

- In high-use recreational parks and working lands, **strategically add native shrub species** where possible to enhance habitat quality for native biodiversity. Wherever possible, replace irrigated turf with shrubs and forbs.

- The cities of Riverside and Jurupa Valley can **incentivize private landowners** and commercial or industrial property owners to **implement native landscaping**, particularly in the strategic priority areas listed above. Incentive programs, outreach, and education, such as through partnership with the Riverside-Corona Resource Conservation District, can promote lawn conversion, rain gardens, green roofs, and other landscaping that utilizes Riversidean sage scrub and forbland plant palettes and improves habitat quality for native wildlife within the urban matrix.

# RESTORATION OF COASTAL SAGE SCRUB

The following recommendations for restoration of coastal sage scrub can be applied to guide Riversidean sage scrub restoration. Riversidean sage scrub is classified as a subassociation of coastal sage scrub, and much of the coastal sage scrub restoration research was conducted in the region.

Coastal sage scrub restoration involves four main stages: *planning, weed management, seeding, and monitoring.* These strategies apply in both urban and natural areas.

*1. Planning* • Restoration should begin with a field assessment of the site to assess soil, slope, existing non-native and native vegetation, site accessibility, and other relevant environmental data (Brooks et al. 2019). Tests of seedbank density or germination responses to invasive plant control can be performed to determine if the native seed bank has been depleted (Allen 2019).

*2. Weed management* • After planning, weed management is needed to control invasive grasses and forbs. An initial dethatch to remove all weeds can be done through mowing, herbicides, or hand pulling (Cione et al. 2002, Allen et al. 2005, Cox and Allen 2008a, Marushia and Allen 2011, Allen et al. 2019, Brooks et al. 2019). Initial dethatching should be followed by a minimum of three years of "grow and kill" weed management two times each year (Brooks et al. 2019, Griswold 2022). Weed removal should occur both in the winter and in spring before seed set.

Solarization, where possible, has been found to be the most successful at removing non-native grasses and may be used in place of "grow and kill" methods (Marushia and Allen 2011, Allen et al. 2019). Solarization involves covering soil with plastic for 40-60 days in the spring (Marushia and Allen 2011, Weathers 2013). However this process depends on sufficient soil moisture and may not always be feasible in Southern California.

Nitrogen deposition from car exhaust pollution can contribute to invasive grass productivity. To combat this, bark mulch with a carbon to nitrogen ratio of 200 or higher can be used for nitrogen immobilization to reduce invasive grass productivity (Allen et al. 1998, Cione et al. 2002).

*3. Seeding* • After three years, when the site has less than ten percent of its original weed cover, the site is ready for seeding. Several studies have found reductions in the native seed bank, indicating the need for seeding as part of restoration efforts (Cione et al. 2002, Cox and Allen 2008b, Brooks et al. 2019). Seed mix should include a diverse mix of key plant functional groups and early successional plants, based upon historical information or adjacent remnant habitats. The Riversidean sage scrub plant palette (Appendix B) provides information on plants including their functional groups (plant type) and historical habitats.

Seeding methods can have a strong impact on restoration outcomes. Imprinting is recommended for small seeded species planted on sites that are flat or gently sloping, while either imprinting or drilling is recommended for seed mixes with many large-seeded species such as perennial grasses and lupines. Hydroseeding should be restricted to steep slopes and rocky areas that are not conducive to imprinting, and when used, the seed mix should bias toward small-seeded species (Montalvo et al. 2002). Planting of seedling plugs can lead to larger shrub establishment; however, due to cost and labor, seeding is often preferred (DeSimone 2011, Allen et al. 2019).

Precipitation can also dramatically impact restoration success and is as important to restoration as weed management (Padgett et al. 2000, Cione et al. 2002, Gillespie and Allen 2004, Cox and Allen 2008a, DeSimone 2011). While annual precipitation cannot be fully predicted, the cyclical nature of El Niño, which brings higher-than-average precipitation to the region, and expert climate forecasts can help inform cost-effective restoration planning to yield the greatest likelihood of success by planting during years of high precipitation (Greenland 1994, Cione et al. 2002, Kimball et al. 2015).

*4. Monitoring* • Following seeding, continued monitoring can inform adaptive management of the site. In particular, it is difficult to predict the extent of invasive species' seedbank at a site, so regular monitoring is needed to understand weed phenology and develop an effective and appropriate weed management plan (Brooks et al. 2019). Monitoring also informs seed mix additions, based on what plants were found to grow from the soil seed bank versus what may need to be seeded (Brooks et al. 2019). Long-term restoration outcomes depend on tracking potential threats, such as the resurgence or invasion of non-native species, and opportunities, such as success of the native seed bank. Detecting and addressing threats early can fortify the project's success, and will be cost-effective in the long term.

Replace irrigated lawns with native shrubs

Plant native shrubs, forbs, and groundcovers in ornamental planting areas in yards

Plant trees next to sports field and rec areas

Plant native shrubs and forbs near remnant RSS patches

**Figure 4.10. Planform graphic highlighting opportunities to introduce native shrubs and near-native tree species within urbanized areas of Riverside.** Trees can be prioritized along buildings, walkways, and roads, and forbs and shrubs can be used as ornamental plantings in yard spaces and medians. (Illustration by Vanessa Lee, SFEI)

Plant trees west and east of houses to reduce heating and cooling costs

To cool homes in summer and warm them in winter, plant only deciduous trees south of homes

Plant near-native trees and native shrubs in medians and adjacent to roads

Plant near-native trees along pedestrian walkways

Riversidean sage scrub (RSS)

120 feet

SANTA ANA RIVER HISTORICAL ECOLOGY 65

# Urban forest:
## *Mitigate extreme heat with strategic tree planting*

Historically, the urban forest habitat zone was largely composed of Riversidean sage scrub and forbland, and tree cover was rare. Today, however, trees are an important element of urban infrastructure. Trees cool cities by reflecting radiation from the sun, cooling through evapotranspiration, and shading surfaces, such as sidewalks and buildings. Thus, in addition to planting Riversidean sage scrub and forbland vegetation in the urban matrix, this zone emphasizes the importance of increasing tree canopy cover to mitigate the urban heat island effect.

We mapped the urban forest habitat zone, shown in Fig. 4.1, using the Trust for Public Land's urban heat island data (The Trust for Public Land 2019). These hot spots within the city would especially benefit from strategic tree planting as a tool to mitigate the effects of extreme heat for residents. Riversidean sage scrub would be an appropriate understory for these areas, as shown in Fig. 4.10.

A minimum 40% canopy cover is needed to achieve significant cooling benefits in urban settings (Lin et al. 2017, Ziter et al. 2019, Ossola et al. 2021). However, even at much lower canopy cover percentages, individual trees can help protect sidewalks and residences from extreme temperatures. Tree cover along pedestrian walkways, bike lanes, parking structures, and on the western and eastern side of buildings, as demonstrated in Fig. 4.10, are especially meaningful for providing cooling benefits.

**Strategically plant trees to provide shade and mitigate urban heat island in areas with high vulnerability.**

- At a landscape scale, select priority tree-planting locations that will mitigate urban heat island effect (see urban forest zones indicated in Fig. 4.1 on page 53).

- At a site-design scale, plant trees where they will generate the greatest cooling benefit for people, as shown in Fig. 4.10.

These strategic locations include planting shade trees on the west, east, and south sides of pedestrian walkways and bike lanes, parking structures, and sports fields. To cool homes and buildings in the summer, it is most important to plant shade trees to the west of the building, and second-most important to plant them to the east. Trees planted to the south of buildings should be deciduous to allow the sun to warm buildings in the winter.

**Prioritize planting native or near-native trees with low to moderate moisture requirements.** Groundwater drawdown and projected increases in drought severity with climate change may lead to increased water scarcity in the future (Hall et al. 2018). Native or near-native trees such as coast live oak (*Quercus agrifolia*), interior live oak (*Q. wislizeni*), Engelmann oak (*Q. engelmannii*), velvet ash (*Fraxinus velutina*), and honey mesquite (*Prosopis glandulosa*) can effectively provide canopy cover in the city, tolerate local drought conditions, and provide food and shelter for various birds, insects, and other native wildlife. Trees may need water during initial establishment, but species should be selected for drought tolerance once established. Some of these species are vulnerable to emerging pests like the shot hole borer (*Euwallacea* spp.) and goldspotted oak borer (*Agrilus auroguttatus*), so monitoring for these will be important to maintain the integrity of an urban forest with oaks or sycamores (L. Larios pers. comm.).

**Plant native shrubs and forbs underneath trees, and in smaller landscaped areas that cannot support trees.** While trees provide the greatest cooling benefit, all vegetation, including shrubs and herbaceous cover, have cooling effects, particularly when vegetation replaces impervious surfaces. Furthermore, planting diverse forms of cover generates structural complexity in the understory, which is beneficial for wildlife and can have attractive aesthetics.

# Conclusion

This pilot study spotlights the immense potential of the Santa Ana River and surrounding areas in and around the City of Riverside to support thriving ecosystems and livable communities with connection to nature and its benefits. The science-based strategies, recommendations, and design guidance presented here are intended to serve as a foundation for future natural resource management and planning efforts, and can be incorporated into Climate Action Plans, Urban Biodiversity Master Plans, Urban Forest Management Plans, habitat restoration projects, and a variety of other planning efforts. Incorporating an understanding of historical habitat patterns and processes can help people develop targeted restoration projects in locations where they are most likely to be successful.

The Santa Ana River functions as a natural ecological and recreational corridor that can be enhanced to provide connections across the landscape for people and wildlife, through the planned Gateway Projects and other future work. Projects along the river and tributary arroyos can enhance infiltration through the underlying coarse alluvial soils and plant native species to help manage the effects of intermittent flooding. Rare wetland features should be prioritized for restoration and protection to enhance habitat diversity on the landscape, and native species, including the Riversidean sage scrub community, can be integrated into urban areas through thoughtful planning. Strategically planted urban trees can protect both wildlife and people against increased extreme heat predicted under climate change. Across all of the habitat types, restoration and management that seeks to enhance connectivity for people and nature will help create a more resilient system now and in the future.

Further research is recommended to more fully investigate historical hydrologic and ecological patterns at the watershed scale and analyze change over time, integrate findings from the historical and urban ecology analyses with future climate change projections, and monitor the success of implementation projects in supporting desired ecological functions and benefits.

**View from Mount Rubidoux, April 1908.** (Courtesy of the Library of Congress)

# REFERENCES CITED

Aerial Information Systems, Inc. 2012. Vegetation - Western Riverside County Update - 2012 [ds1196]. Western Riverside County Regional Conservation Authority.

Allen, E. B. 2019. Long-term prospects for restoration of coastal sage scrub: invasive species, nitrogen deposition, and novel ecosystems. Page 18. Gen. Tech. Rep., U.S. Department of Agriculture, Forest Service, Pacific Southwest Research Station, Albany, CA.

Allen, E. B., R. D. Cox, T. Tennant, S. N. Kee, and D. H. Deutschman. 2005. Landscape restoration in southern California forblands: response of abandoned farmland to invasive annual grass control. Israel Journal of Plant Sciences 53:237–245.

Allen, E. B., C. McDonald, and B. E. Hilbig. 2019. Long-term prospects for restoration of coastal sage scrub: invasive species, nitrogen deposition, and novel ecosystems. Pages 1–18 Proceedings of the chaparral restoration workshop, California. Gen. Tech. Rep. PSW-GTR-265. Albany, CA: US Department of Agriculture, Forest Service, Pacific Southwest Research Station: 1-18.

Allen, E. B., P. E. Padgett, A. Bytnerowicz, and R. Minnich. 1998. Nitrogen Deposition Effects on Coastal Sage Vegetation of Southern California. Page 131 Proceedings of the International Symposium on Air Pollution and Climate Change Effects on Forest Ecosystems, February 5-9, 1996, Riverside, California. US Department of Agriculture, Forest Service, Pacific Southwest Research Station.

Anderson, K. 2005. Tending the Wild: Native American Knowledge and the Management of California's Natural Resources. University of California Press.

Bailey, A. M., H. K. Ober, B. E. Reichert, and R. A. McCleery. 2019. Canopy Cover Shapes Bat Diversity across an Urban and Agricultural Landscape Mosaic. Environmental Conservation 46:193–200.

Baum, Kristen A., Kyle J. Haynes, Forrest P. Dillemuth, and James T. Cronin. 2004. The Matrix Enhances the Effectiveness of Corridors and Stepping Stones. Ecology 85: 2671–76.

Barbour, M., T. Keeler-Wolf, and A. A. Schoenherr. 2007. Sage Scrub. Pages 208–228 Terrestrial Vegetation of California. Univ of California Press.

Bean, L. J. 1972. Mukat's people; the Cahuilla Indians of southern California. Berkeley, University of California Press.

Bean, L. J., and K. S. Saubel. 1972. Temalpakh: Cahuilla Indian Knowledge and Usage of Plants. Malki Museum Press.

Bean, L. J., and F. C. Shipek. 1978. Luiseño. Pages 550–563 Handbook of North American Indians. Smithsonian Institution, Washington, D.C.

Bean, L.J., and C.R. Smith. 1978. Serrano. In Handbook of North American Indians. Washington D.C.: Smithsonian Institution. Washington, D.C.

Bell, G. P. 1998. Ecology and management of Arundo donax, and approaches to riparian habitat restoration in southern California. In Brock, J. H., Wade, M., Pysek, P., and Green, D. (Eds.): Plant Invasions: Studies from North America and Europe. Blackhuys Publishers, Leiden, The Netherlands, pp. 103-113.

Beller, E., R. Grossinger, M. Salomon, S. Dark, E. Stein, B. Orr, P. Downs, T. Longcore, G. Coffman, A. Whipple, R. Askevold, B. Stanford, and J. Beagle. 2011. Historical ecology of the lower Santa Clara River, Ventura River, and Oxnard Plain: an analysis of terrestrial, riverine, and coastal habitats. 641, San Francisco Estuary Institute, Oakland.

Beninde, J., M. Veith, and A. Hochkirch. 2015. Biodiversity in cities needs space: a meta-analysis of factors determining intra-urban biodiversity variation. Ecology Letters 18:581–592.

Blake, W. P. 1857. Explorations and Surveys for a Railroad Route From the Mississippi River to the Pacific Ocean, vol. 5. Geological Report, Washington, D.C.

Bolton, H. E. 1930. Anza's California Expeditions. University of California Press, Berkeley, CA.

Brooks, T., M. Griswold, T. Longcore, and M. Riedel-Lehrke. 2019. Habitat Restoration and Enhancement Plan Update. Land IQ.

Brown, J., and J. Boyd. 1922. History of San Bernardino and Riverside Counties: With Selected Biography of Actors and Witnesses of the Period of Growth and Achievement... Western Historical Association.

CalFire. 2017. Fire Perimeter. Fire and Resource Assessment Program.

Calflora. 2021. Calflora. https://www.calflora.org/.

California Native Plant Society. 2021. CNPS Rare Plant Inventory. https://rareplants.cnps.org/Home/Index.

Cao, X., A. Onishi, J. Chen, and H. Imura. 2010. Quantifying the cool island intensity of urban parks using ASTER and IKONOS data. Landscape and urban planning 96:224–231.

Chang, C.-R., M.-H. Li, and S.-D. Chang. 2007. A preliminary study on the local cool-island intensity of Taipei city parks. Landscape and urban planning 80:386–395.

Cione, N. K., P. E. Padgett, and E. B. Allen. 2002. Restoration of a native shrubland impacted by exotic grasses, frequent fire, and nitrogen deposition in southern California. Restoration Ecology 10:376–384.

City of Riverside Public Utilities. 2020. Groundwater Atlas. https://riversideca.gov/utilities/sites/riversideca.gov.utilities/files/pdf/2018_Riverside_Atlas.pdf.

Consortium of California Herbaria. 2022. CCH2 data portal.

Cox, R. D., and E. B. Allen. 2008a. Stability of exotic annual grasses following restoration efforts in southern California coastal sage scrub. Journal of Applied Ecology 45:495–504.

Cox, R. D., and E. B. Allen. 2008b. Composition of soil seed banks in southern California coastal sage scrub and adjacent exotic grassland. Plant Ecology 198:37–46.

DeSimone, S. A. 2011. Balancing active and passive restoration in a nonchemical, research-based approach to coastal sage scrub restoration in southern California. Ecological Restoration 29:45–51.

Dunlap, J. C. 1886. Map showing the lands of the East Riverside Land Co, San Bernardino Co, Cal. Courtesy of San Bernardino County Department of Public Works.

EarthDefine. 2016. 1-meter resolution US land cover. EarthDefine.

EcoAdapt. 2016. Sage Scrub Habitats Climate Change Vulnerability, Adaptation Strategies, and Management Implications in Southern California National Forests.

Elmore, A. J., S. J. Manning, J. F. Mustard, and J. M. Craine. 2006. Decline in alkali meadow vegetation cover in California: the effects of groundwater extraction and drought. Journal of Applied Ecology 43:770–779.

Environmental Law Institute. 2003. Conservation Thresholds for Land Use Planners. Environmental Law Institute, Washington, D.C.

Esri, DigitalGlobe, GeoEye, i-cubed, USDA, FSA, USGS, AEX, Getmapping, Aerogrid, IGN, IGP, swissopo, and GIS User Community. 2022, September 21. "World imagery" [basemap]. Esri.

Evans, M. J., S. C. Banks, D. A. Driscoll, A. J. Hicks, B. A. Melbourne, and K. F. Davies. 2017. Short- and Long-Term Effects of Habitat Fragmentation Differ but Are Predicted by Response to the Matrix. Ecology 98: 807–19.

Fairchild Aerial Surveys. 1931. Flight C-1740. Courtesy University of California Santa Barbara Library.

Font, P., and A. K. Brown. 2011. With Anza to California, 1775-1776: the journal of Pedro Font, O.F.M. Page (A. K. Brown, Ed.). The Arthur H. Clark Comapny, an imprint of the University of Oklahoma Press, Norman, Oklahoma.

Friggens, M. M., M. V. Warwell, J. C. Chambers, and S. G. Kitchen. 2012. Modeling and Predicting Vegetation Response of Western USA Grasslands, Shrublands, and Deserts to Climate Change. Page Climate change in grasslands, shrublands, and deserts of the interior American West. USDA Forest Service.

GEI Consultants, Inc. and CWE. 2020. Assessing Homelessness Impacts on Water Quality, Riparian and Aquatic Habitat in Upper Santa Ana River Watershed. Submitted to: Santa Ana Watershed Project Authority.

Gillespie, I. G., and E. B. Allen. 2004. Fire and competition in a southern California grassland: impacts on the rare forb Erodium macrophyllum. Journal of Applied Ecology 41:643–652.

Goldworthy, and Higbie. 1871. Map of 10 Acre Lots The Property of the S.C.C. Association Situated on the Jurupa Rancho San Bernadino. San Bernadino, CA. Courtesy of San Bernardino County Department of Public Works.

Google Earth. 2017-20. "Riverside." 33°58'43.17"N and 117°24'11.21"W. Google Earth.

GreenInfo Network. 2020. California Protected Areas Database. GreenInfo Network.

Greenland, D. 1994. El Niño and long-term ecological research (LTER) sites. LTER Network Office Publication.

Greves, J. P. 1876. "History of Riverside." Journal of the Riverside Historical Society, Number Six, 2002. Inland Printworks: Riverside, CA.

Griswold, M. 2022, April 19. Coastal Sage Scrub Restoration. (Personal Communication).

Grossinger, R. M., C. J. Striplen, R. A. Askevold, E. Brewster, and E. E. Beller. 2007. Historical landscape ecology of an urbanized California valley: wetlands and woodlands in the Santa Clara Valley. Landscape Ecology 22:103–120.

Gunn, C. W. 1885. On Six Species of Hummingbirds of the Pacific Slope. Ornithologist 10:26.

Hall, A., N. Berg, and K. Reich. 2018. Los Angeles Region Report. Page 97. University of California, Los Angeles.

Hall, H. M. 1895. Nesting of the Western Yellow-Throat. The Nidiologist 1:137.

Hall, H. M. 1905. Hall, Harvey M. Field Book #3. Courtesy of The Jepson Herbarium, University of California Berkeley.

Hall, W. M. H. California Office of State. 1888a. Irrigation in California (Southern): The Field, Water-Supply, and Works, Organization and Operation in San Diego, San Bernardino, and Los Angeles Counties: The Second Part of the Report of the State Engineer of California on Irrigation and the Irrigation Question. J.D. Young, Superintendent State Print.

Hall, W. M. H. 1888b. Detail Irrigation Map Riverside Sheet. California State Engineering Department.

Hancock, H. 1853a. Field Notes of the Boundary Lines of Townships 1 + 2 South Range 5 West. U.S. Surveyor General. Courtesy of Bureau of Land Management.

Hancock, H. 1853b. Field Notes of the Exterior Lines of Townships 1, 2, + 3 South Ranges 5 + 6 West San Bernardino Meridian. U.S. Surveyor General. Courtesy of Bureau of Land Management.

Hancock, H. 1858. Field Notes of the Final Survey of Rancho Jurupa Louis Rubideau Confirmee. U.S. Surveyor General. Courtesy of Bureau of Land Management.

Hanes, T. L., R. D. Friesen, and K. Keane. 1989. Alluvial Scrub Vegetation in Coastal Southern California. Page USDA Forest Service General Technical Report. Davis, CA.

Hayes, B. 1863. Pioneer notes from the diaries of Judge Benjamin Hayes, 1849-1875. Page 257. Courtesy of Library of Congress.

Heaviside, C., H. Macintyre, and S. Vardoulakis. 2017. "The Urban Heat Island: Implications for Health in a Changing Environment." Current Environmental Health Reports 4: 296–305.

Heizer, R. F. 1978. Handbook of North American Indians. Smithsonian Institution, Washington.

Hesselbarth, M. K., M. Sciaini, K. A. With, K. Wiegand, and J. Nowosad. 2019. "Landscapemetrics: An Open-Source R Tool to Calculate Landscape Metrics." Ecography 42 (10): 1648–57.

Horne, M. C., and D. P. McDougall. 2007. Cultural Resources Study for the City of Riverside General Plan 2025 Update Program EIR And. Hemet, California: Applied EarthWorks, Inc.

ICF. 2020. Upper Santa Ana River habitat conservation plan. October 2020 stakeholder draft. Prepared for San Bernardino Valley Municipal Water District.

Jin, Y., M. L. Goulden, N. Faivre, S. Veraverbeke, F. Sun, A. Hall, M. S. Hand, S. Hook, and J. T. Randerson. 2015. "Identification of Two Distinct Fire Regimes in Southern California: Implications for Economic Impact and Future Change." Environmental Research Letters 10 (9): 094005.

Kimball, S., M. Lulow, Q. Sorenson, K. Balazs, Y. Fang, S. J. Davis, M. O'Connell, and T. E. Huxman. 2015. Cost-Effective Ecological Restoration. Restoration Ecology 23:800–810.

Kelly, M., B. Allen-Diaz, and N. Kobzina. 2005. Digitization of a historic dataset: the Wieslander California vegetation type mapping project. Madroño 52:191–201.

Koen, E. L., J. Bowman, C. Sadowski, and A. A. Walpole. 2014. Landscape connectivity for wildlife: development and validation of multispecies linkage maps. Methods in Ecology and Evolution 5:626–633.

Larios, L. 2023, April 21. (Personal Communication).

Laub, B. G., J. Detlor, and D. L. Keller. 2020. Determining factors of cottonwood planting survival in a desert river restoration project. Restoration Ecology 28:A24–A34.

Lech, S. 2004. Along the old road: a history of the portion of Southern California that became Riverside County, 1772-1893. Published by the author.

Lin, P., S. S. Y. Lau, H. Qin, and Z. Gou. 2017. Effects of urban planning indicators on urban heat island: a case study of pocket parks in high-rise high-density environment. Landscape and Urban Planning 168:48–60.

Lindley, W. and J. P. Widney. 1888. California of the south: its physical geography, climate, resources, routes of travel, and health-resorts. New York: D. Appleton and Company.

Lu, R., K. C. Schiff, and K. D. Stolzenbach. 2003. Nitrogen deposition on coastal watersheds in the Los Angeles region:9.

Magness, E. 1899. With rainbow trout in Southern California. Forest and Stream vol. 52.

Malanson, George P. 2003. Dispersal across Continuous and Binary Representations of Landscapes. Ecological Modelling 169: 17–24.

Marushia, R. G., and E. B. Allen. 2011. Control of exotic annual grasses to restore native forbs in abandoned agricultural land. Restoration Ecology 19:45–54.

Matteson, K. C., and G. A. Langellotto. 2010. Determinates of inner city butterfly and bee species richness. Urban Ecosystems 13:333–347.

May, M. R., M. C. Provance, A. C. Sanders, N. C. Ellstrand, and J. Ross-Ibarra. 2009. A Pleistocene Clone of Palmer's Oak Persisting in Southern California. PLoS One 4: e8346.

McCawley, W. 1996. The First Angelinos: The Gabrielino Indians of Los Angeles. Ballena Press.

McLain, R. B. 1899. Critical Notes on a Collection of Reptiles From the Western United States. Contributions to North American Herpetology. Wheeling, West Virginia.

McRae, B. 2012. Barrier Mapper Connectivity Analysis Software. The Nature Conservancy, Seattle, WA.

McRae, B. H., B. G. Dickson, T. H. Keitt, and V. B. Shah. 2008. Using circuit theory to model connectivity in ecology, evolution, and conservation. Ecology 89:2712–2724.

McRae, B. H., V. B. Shah, and T. K. Mohapatra. 2013. Circuitscape 4 User Guide. The Nature Conservancy.

McRae, B., and D. Kavanagh. 2011. Linkage Mapper Connectivity Analysis Software. The Nature Conservancy, Seattle, WA.

Mendenhall, W. C. 1905. Hydrology of the San Bernardino Valley. Page Water-supply And Irrigation Paper 142. Department of the Interior United States Geological Survey.

Miller and Newman. 1876. Rubidoux Rancho especially showing lands sold by Louis Rubidoux. Courtesy of San Bernardino County Department of Public Works.

Milliken, R., J. Johnson, D. Earle, N. Smith, P. Mikkelsen, P. Brandy, and J. King. 2010. Introduction to the Contact-Period Native California Community Distribution Model. [Digital map of ethnographic regions.]. Far Western Anthropological Research Group, Inc.

Minnich, R. A. 2008. California's Fading Wildflowers: Lost Legacy and Biological Invasions.

Minnich, R., and R. Dezzani. 1998. Historical decline of coastal sage scrub in the Riverside-Perris Plain, California. Western Birds 29.

Minto, W. M. 1878. Field Notes of the Subdivision and Exterior Lines of Township 2 South, Range 5 West San Bernardino Meridian, California. U.S. Surveyor General. Courtesy of Bureau of Land Management.

Montalvo, A. M., P. A. McMillan, and E. B. Allen. 2002. The relative importance of seeding method, soil ripping, and soil variables on seeding success. Restoration Ecology 10:52–67.

National Hydrography Dataset. 2019. U.S. Geological Survey.

Nelson, J. W., R. L. Pendleton, J. E. Dunn, A. T. Strahorn, and E. B. Watson. 1915. Soil Map California Riverside Area. U.S. Department of Agriculture Bureau of Soils.

Nelson, J. W., R. L. Pendleton, J. E. Dunn, A. T. Strahorn, and E. B. Watson. 1917. Soil Survey of the Riverside Area, California. U.S. Department of Agriculture, Washington.

Nordhoff, C. 1873. California: for health, pleasure, and residence. A book for travelers and settlers. Harper & Brothers, New York.

Ossola, A., G. D. Jenerette, A. McGrath, W. Chow, L. Hughes, and M. R. Leishman. 2021. Small vegetated patches greatly reduce urban surface temperature during a summer heatwave in Adelaide, Australia. Landscape and Urban Planning 209:104046.

Padgett, P. E., S. N. Kee, and E. B. Allen. 2000. The effects of irrigation on revegetation of semi-arid coastal sage scrub in southern California. Environmental management 26:427.

Patterson, T. 1996. A colony for California. The Museum Press of the Riverside Museum Associates, Riverside.

Placeworks. 2018, March 18. Parkway and Open Space Plan. Coastal Conservancy.

Reed, F. M. 1916. Catalog of the plants of Riverside and vicinity. Riverside, California. Courtesy of HathiTrust.

Roberts, F. M., S. D. White, A. C. Sanders, D. E. Bramlet, and S. Boyd. 2004. The Vascular Plants of Western Riverside County, California. F.M. Roberts Publications

Robinson, W. W. 1957. The Story of Riverside County. Title Insurance and Trust Company.

Rutter, C. 1896. Notes of Freshwater Fishes of the Pacific Slope of North America. Proceedings of the California Academy of Sciences, 2nd series, vol. 6. California Academy of Sciences, San Francisco.

Safran, S., S. Baumgarten, E. Beller, J. Crooks, R. Grossinger, J. Lorda, T. Longcore, D. Bram, S. Dark, E. Stein, and T. McIntosh. 2017. Tijuana River Valley Historical Ecology Investigation. San Francisco Estuary Institute.

San Bernardino Valley Municipal Water District, and Western Municipal Water District. 2004. Appendix A: Surface water hydrology. Santa Ana River water rights applications for supplemental water supply draft environmental impact report.

Sanborn, K. 1904. Map Showing Riverside Water Co's Rubidoux Pumping Plant. Riverside, California. Courtesy of City of Riverside Public Works Department.

Saubel, K. S., and E. Elliot. 2004. Isill Héqwas Wáxish = A Dried Coyote's Tail. Banning, CA: Malki Museum Press.

SAWPA. 2019. One Water One Watershed Plan Update 2018: Moving forward together. Santa Ana Watershed Project Authority.

Snyder, J. O. 1908. Description Of Pantosteus Santa-anae a New Species of Fish From the Santa Ana River, California. Pages 33–34 Proceedings of the U.S. National Museum, vol. 34.

Southern California Association of Governments. 2019. 2019 Annual Land Use. Southern California Association of Governments.

Spotswood, E., R. Grossinger, S. Hagerty, M. Bazo, M. Benjamin, E. Beller, L. Grenier, and R. Askevold. 2019. Making Nature's City: A Science-Based Framework for Building Urban Biodiversity. San Francisco Estuary Institute. Publication.

Stillwater Sciences. 2007. Analysis of Riparian Vegetation Dynamics for the Lower Santa Clara River and Major Tributaries, Ventura County, California. Santa Clara River Parkway Floodplain Restoration Feasibility Study. Prepared by Stillwater Sciences for the California State Coastal Conservancy and the Santa Clara River Trustee Council.

Street, R. S. 2004. Beasts of the field: a narrative history of California farmworkers, 1769-1913. Stanford University Press.

Sugihara, N. G., editor. 2006. Fire in California's ecosystems. University of California Press, Berkeley.

Swarth, H. S. 1908a. Section 2: Riverside area, southern California 1908. Pages 113–115 Field Notes: Swarth H.S. 1908, v1678. Courtesy of Museum of Vertebrate Zoology, University of California Berkeley.

Swarth, H. S. 1908b. Section 4: Jurupa Mts. area, southern California 1908. Pages 144–146 Field Notes: Swarth H.S. 1908, v1678. Courtesy of Museum of Vertebrate Zoology, University of California Berkeley.

Taha, H., and T. Freed. 2015. Creating and Mapping an Urban Heat Island Index for California. Report Prepared by Altostratus Inc., Contract, 13–001.

The Jepson Herbarium. 2021. California Flora, Jepson eFlora Main Page. https://ucjeps.berkeley.edu/eflora/.

The Trust for Public Land. 2019. Urban Heat Island Severity for U.S. cities - 2019. U.S.G.S.

Twogood, F. D. 1897. Rhopalocera of Riverside, California. Entomological news, and proceedings of the Entomological Section of the Academy of Natural Sciences of Philadelphia:29–32.

Tyson, P. T., P. F. Smith, and R. S. Williamson. 1851. Geology and Industrial Resources of California. W. Minifie & Company, Baltimore.

Unknown. 1872. Field Notes of the Final Survey of Rancho Jurupa Louis Rubideau Confirmee. U.S. Surveyor General. Courtesy of Bureau of Land Management.

U.S. Census Bureau. 2019. 2019 TIGER/Line Shapefiles. U.S. Census Bureau.

USDC (U.S District Court). 1854-58. "Rancho Jurupa." Land Case Map F-1247. Courtesy of The Bancroft Library, UC Berkeley.

USGS (U.S. Geological Survey). 1901. Riverside Quadrangle, California: 15-minute series (Topographic).

U.S. Surveyor General's Office. 1878. Plat of the Jurupa Rancho finally confirmed to Abel Stearns. CS 986. Courtesy of San Bernardino County Surveyor.

Wang, Y.-C., J.-K. Shen, and W.-N. Xiang. 2018. Ecosystem service of green infrastructure for adaptation to urban growth: function and configuration. Ecosystem Health and Sustainability 4:132–143.

Weathers, K. A. 2013. The role of invasive Erodium species in restoration of Coastal Sage Scrub communities and techniques for control. University of California, Riverside.

Western Riverside County Regional Conservation Authority. 2003. Western Riverside County Multiple Species Habitat Conservation Plan. https://rctlma.org/Portals/0/mshcp/volume2/SectionB.html.

Williamson, R. S. 1856. The Valley of San Bernardino. Pages 81–82 Reports of Explorations and Surveys to Ascertain the Most Practicable and Economical Route for a Railroad from the Mississippi River to the Pacific Ocean, vol. 5. War Department, Washington.

Wood, E. M., and S. Esaian. 2020. The importance of street trees to urban avifauna. Ecological Applications 30:e02149.

Yarnell, S. M., E. D. Stein, J. A. Webb, T. Grantham, R. A. Lusardi, J. Zimmerman, R. A. Peek, B. A. Lane, J. Howard, and S. Sandoval-Solis. 2020. A functional flows approach to selecting ecologically relevant flow metrics for environmental flow applications. River Research and Applications 36:318–324.

Ziter, C. D., E. J. Pedersen, C. J. Kucharik, and M. G. Turner. 2019. Scale-dependent interactions between tree canopy cover and impervious surfaces reduce daytime urban heat during summer. Proceedings of the National Academy of Sciences 116:7575–7580.

# APPENDIX A: Detailed Methods

## Urban ecology assessment

To assess the present-day landscape within the study area, we evaluated each of seven elements known to be important to biodiversity support (Spotswood et al. 2019). These elements and the methods we used to analyze them are described briefly here.

**Patch Size**—the area of each discrete greenspace in a city—is one of the two core drivers of urban biodiversity. Patches can range from smaller neighborhood parks to golf courses, cemeteries, and large city parks, as well as natural spaces such as forests and lakes. Larger patches generally support greater biodiversity because they contain more kinds of habitats and provide more resources than smaller patches. Habitat patches that support biodiversity in cities are generally protected areas of greenspace greater than 2 ac (0.8 ha) in size, but can also include informal or privately-managed greenspaces, such as backyards, school yards, or vacant lots. While 2 ac is the smallest area required to be considered a viable patch, biodiversity declines rapidly when greenspaces are smaller than 10 ac (4 ha) in size. Large patches (greater than 130 ac [53 ha] in size) are considered regional biodiversity hubs and can host species that are area-sensitive and intolerant of urban environments (Table A1).

To identify habitat patches within the study area, we identified contiguous areas of vegetation making up at least 2 ac by analyzing EarthDefine (2016) land cover data using the "landscapemetrics" R package (Hesselbarth et al. 2019). Long, linear stretches of vegetation, such as strings of backyards and median strips, were filtered out, as they do not provide core habitat and are better characterized as matrix quality, a separate landscape element that also supports biodiversity within the urban matrix. This filtering was done by calculating each patch's Contiguity Index, which is a metric that characterizes each patch's shape and spatial connectedness. Only patches with a contiguity index above the 75th percentile for the study area were selected. These patches were then manually reviewed and updated based on 2022 aerial imagery (Esri et al. 2022), to correct for changes in land cover since 2016. We classified these protected open spaces into three categories based on the size thresholds that are meaningful for evaluating their potential for supporting biodiversity: greater than 2 ac, 10 ac, and 130 ac.

We then overlapped these patches with protected areas databases, the California Protected Areas Database (GreenInfo Network 2020) and the Southern California Association of Governments (SCAG) Annual Land Use Dataset (SCAG 2019), to distinguish between formally protected habitat and informal, unprotected, or privately-managed patches. The types of

Table A1. Larger habitat patches yield more benefits for both urban biodiversity and urban residents. In general, this table outlines the benefits associated with habitat patches above each size threshold.

| Patch size | Benefits | Source |
|---|---|---|
| > 2 acres | Supports many urban-adapted species | Beninde et al. (2015), Spotswood et al. (2019) |
| > 5 acres | Regulates local microclimates and decreases urban heat island effects | Chang et al. (2007), Cao et al. (2010), Wang et al. (2018) |
| > 10 acres | Supports urban-sensitive species | Beninde et al. (2015) |
| > 30 acres | Generates greater, more consistent cooling effects on the surrounding urban landscape | Chang et al. (2007), Wang et al. (2018) |
| > 130 acres | Supports area-sensitive or forest-interior species | Beninde et al. (2015) |

opportunities and interventions to expand and enhance habitat patches in the study area differ, based on their protection status. Government entities have more agency to strategically manage and restore core habitat on sites that are publicly protected. While unprotected or privately managed sites can also play a key role in supporting urban biodiversity, they will require more coordination, education, strategic acquisition, and, potentially, regulation to enhance habitat quality and ensure long-term preservation.

**Connections** are linear vegetated features that facilitate the movement of plants and animals between habitat patches. Contiguous stretches of vegetation linking wider greenspaces, such as green corridors along waterways and right-of-ways, form some of the most effective connections in cities. In the absence of continuous corridors, "stepping stones" of matrix habitat, such as closely-spaced pocket parks, backyards, vacant lots, and green roofs can increase the ability of species to move between patches.

To identify the pathways that wildlife are most likely to use when traveling across the landscape, we modeled connectivity within the study area using Circuitscape (McRae et al. 2008, 2013). We applied a generalized, species-agnostic approach developed by Koen et al. (2014). We reclassified land cover developed by EarthDefine (2016) into a "resistance" surface that represents most species' preferences for vegetated areas, such as shrub cover or tree canopy cover, over highly impervious developed sites. Since this approach is meant to represent wildlife and plant movement generally, it does not account for species-specific habitat requirements. The landscape connectivity results generated by the Circuitscape model were then used to map important barriers to connectivity using the Linkage Pathways tool (McRae and Kavanagh 2011, McRae 2012).

**Matrix Quality** refers to how well urban areas support biodiversity in between patches and corridors. Areas with more street trees, bioretention areas, green roofs, and backyard gardens are better able to support native plants and animals. While individual habitat elements in the matrix are often too small to support large wildlife populations themselves, they can support wildlife movement and foraging in cities. Matrix quality improvements can be made around patches to increase the effective patch size, along connections to increase the effective corridor width, between patches to increase connectivity, or clustered to form habitat complexes.

To assess matrix quality, we evaluated vegetation cover and heat risk across the study area. Using 1-meter resolution land cover data developed by EarthDefine (2016), we calculated the percent of total area that is composed of both shrub cover and tree canopy cover within each U.S. census block. To identify areas of extreme heat risk, we used the Trust for Public Land's (TPL) map of heat severity for cities across the United States (Trust for Public Land 2019). TPL analyzed Landsat 8 satellite imagery from the summers of 2018 and 2019 to assess which areas within a city are hotter than average, relative to the city as a whole.

**Habitat Diversity** refers to the type, number and spatial arrangement of habitats within the urban area. Creative landscape planning and design that reflects the scale, complexity, arrangement, and diversity of habitats that were historically present on the landscape increases the total resources available and better supports urban biodiversity. When planning for habitat diversity, it is important to both promote coherence and heterogeneity at the city scale and to mimic the spatial complexity, vertical structure, and physical features of individual habitats at the site scale. Protecting and augmenting rare native habitats in cities can be particularly beneficial for habitat specialists, which may be especially vulnerable to habitat loss.

**Native Vegetation** includes plant species that have a long evolutionary history in a particular location. Native plants support the native wildlife with which they have co-evolved. For example, many insects have developed specialized relationships with native host plants. Native plants can bolster the entire food web by supporting the presence of these specialized local insects, which can, in turn, be a food resource for other wildlife. In addition to providing wildlife habitat, the use of native species in urban landscaping can also reduce water usage and maintenance costs. Selecting native species and communities that are likely to be tolerant of future climate conditions, particularly for long-lived plants and trees, can create a climate-adaptive native or largely native plant palette.

The Western Riverside County Regional Conservation Authority commissioned the update of vegetation mapping within the footprint of the Western Riverside County Multi-Species Habitat Conservation Plan (Aerial Information Systems, Inc. 2012). While the mapping was completed in 2005 and updated in 2012, it currently presents the most detailed mapping of vegetation communities within the study area. The vegetation mapping includes information both on the diversity of habitats and native plants that are present across the study area. The mapping also identifies areas where invasive plants are particularly dominant and have displaced native vegetation.

**Special Resources** are components of an ecosystem that can provide disproportionate benefits to wildlife. Special resources can help animals meet their needs for food, shelter, or water during all or part of the year. For example, large trees and well-designed urban water bodies serve as hubs for local biodiversity. Trees with cavities for nesting birds and woody debris piles for reptiles and insects, which are typically removed in urban environments, can support specialists and increase biodiversity in otherwise resource-limited areas. Artificial structures such as nest boxes and bat caves can provide critical features in small spaces.

Particularly rare or productive landscape features within the study area include riparian plant communities, aquatic resources, and remnant scrub habitat. We evaluated the presence of these special resources using the Western Riverside County vegetation map (Aerial Information Systems, Inc. 2012) and land cover mapping (EarthDefine 2016).

**Management** includes interventions or practices adopted by land or facility managers that create changes in the landscape. Biodiversity-friendly management actions include reducing pesticide and herbicide use, minimizing disturbance to sensitive wildlife areas, removing invasive species especially from yards on the urban wildland interface, limiting the impacts of domestic cats and dogs, reducing light and noise pollution, and regulating human activity to reduce conflict with wildlife. Design actions such as fitting buildings with bird-safe windows and creating wildlife underpasses and overpasses are also essential to creating a more wildlife-friendly built environment.

To identify key management challenges within the study area, we spoke with local natural resource managers and stakeholders and reviewed regional policy and planning documents outlined in Chapter 1.

# Habitat zone mapping

Habitat zones were mapped by overlaying numerous geospatial datasets to understand historical and current conditions and assess the opportunities and constraints for habitat improvement.

The central dataset informing habitat zone development is the historical ecology map. The historical ecology results describe the distribution of plant communities within the study area prior to urban development. This historical distribution reflects underlying gradients of temperature, topography, soils, and water availability, which in turn inform potential present-day restoration opportunities.

While historical ecology is a helpful guide to inform restoration potential, land use and physical conditions have changed extensively over the past two centuries. We used several contemporary sources to map habitat zones along two dimensions: the habitat type that is appropriate within a given zone and the level of land use intensity that poses a unique set of opportunities and constraints for supporting urban biodiversity (Table A2).

To map appropriate habitat types, we used the 2012 Western Riverside County Multi-Species Habitat Conservation Plan (MSHCP) vegetation mapping to identify where remnant habitats remain intact on the landscape (Aerial Information Systems, Inc. 2012). A contemporary map of waterways and waterbodies in the study area further supported our mapping of arroyo habitat (National Hydrography Dataset 2019). We then conducted a detailed review of the study area using world imagery (Esri et al. 2022), supplemented by Google Earth imagery (Google Earth 2017-20). These aerial and satellite images supported our delineation of habitat types.

While the natural structure of upland regions of the study area was generally short-statured and treeless, trees play an important role in provisioning resources, managing heat and stormwater, and improving health and wellbeing for people in cities. Therefore, we also introduced a new habitat zone, which we call the urban forest. We mapped the urban forest habitat zone within highly urbanized regions of the study area that experience significant urban heat island effects (Trust for Public Land 2019).

To map land use intensity, we used EarthDefine's land cover mapping of impervious cover to identify the zones of highest land use intensity (EarthDefine 2016). We used the MSHCP vegetation map to identify agricultural lands (Aerial Information Systems, Inc. 2012), and park datasets to identify high-use recreational open spaces (SCAG 2019, GreenInfo Network 2020). Where detailed land use information was not available, reviewing world imagery further supported our interpretation of land use intensity (Esri et al. 2022).

Table A2. Descriptions of the land use intensity classification used to map habitat zones.

| Land-use intensity class | Description | Examples |
| --- | --- | --- |
| Low | Habitat patches with high native vegetation cover, low-impact recreational uses, and low impervious cover | Natural parks and open space |
| Moderate | Large parks and open spaces that have a high level of modification and recreational human use, high non-native vegetation cover, and a moderate-to-low amount of impervious cover | Golf courses, sports fields, cemeteries, and agricultural fields |
| High | Areas with high impervious cover, low vegetation cover, and high-impact land uses | Residential, commercial, and industrial development |

# APPENDIX B: Plant Palettes

## Plant palette methods

The plant list was compiled by extracting the names of all species historically found within our study area from Consortium of California Herbaria data (Consortium of California Herbaria 2022). We included all geolocated species with location uncertainty buffers overlapping our study area, and all non-geolocated data with locality descriptions suggesting they occurred in our study area. Only records prior to 1950 were included. Plants were then assigned habitat associations using habitat data from The Vascular Plants of Western Riverside County, California checklist; CalFlora; CNPS Rare Plant Inventory; and Jepson eFlora (Roberts et al. 2004, Calflora 2021, California Native Plant Society 2021, The Jepson Herbarium 2021). Plants listed as occurring in river washes were coded as occurring in alluvial scrub, as river washes were presumed largely barren in our habitat mapping. Dominant species are highlighted.

Several additional criteria were used to further filter plant palettes for each habitat area. Chosen characteristics included low water requirements, availability in nurseries, ease of care, and wildlife support. Consideration of cultural uses was also included, with knowledge provided by Aaron Saubel of the Malki Museum. Along with these characteristics, provision of year round bloom time and perennial herbs were prioritized to create final plant palettes for each historical habitat type. Information about common name, plant type, wildlife support, flower color, bloom time, soil, drainage, and ease of care comes from the Calscape (www.calscape.org) and/or Calflora (www.calflora.org) databases.

Because the Los Angeles region is projected to become hotter and experience more frequent droughts with climate change, practitioners should emphasize the use of more xeric-adapted plants (i.e., those with desert affinities) in locations most vulnerable to extreme heat.

# Riversidean sage scrub plant palette

| Scientific | Common | Plant Type | Wildlife Support | Cultural Use | Flowers | Bloom Time | Soil | Drainage | Ease of Care |
|---|---|---|---|---|---|---|---|---|---|
| *Aristida purpurea* | Purple Threeawn; Red Threeawn | Grasses | | | Cream, Purple, Red, Brown | | Tolerates a variety of soils | Fast, Medium | Very Easy |
| *Artemisia californica* | California Sagebrush | Shrub | California Gnatcatcher, Quail, various other birds, insects, butterflies | Used medicinally and for smudging | Cream, White, Yellow | Apr-Oct | Usually found on very dry slopes or sandy soil with low nutrient content, although it is also said to tolerate clay | Fast, Medium, Slow | Very Easy |
| *Artemisia palmeri* | San Diego Sagewort | Shrub | Numerous birds are attracted to the seeds, including quail, thrashers, towhees and finches | | Cream, White, Yellow | May-Sep | Typically sandstone or sandy soil | Fast, Medium | Moderately Easy |
| *Calystegia macrostegia* | Island False Bindweed | Perennial herb, Shrub, Vine | Butterflies, bees, moths | | Pink, White | Feb-Jul | Typically sandy and/or rocky soil | Fast, Medium | Very Easy |
| *Castilleja exserta* | Indian Paintbrush; Exserted Indian Paintbrush | Annual Herb | Bees and butterflies. This is a crucial host plant for the Bay Checkerspot butterfly, a threatened species in California. Beneficial for the Leanira Checkerspot and Chalcedon Checkerspot butterflies. | | Lavender, Pink, Purple | Mar-Jun | Tolerant of sand and clay | Fast, Medium, Slow | Moderately Easy |
| *Chaenactis glabriuscula* | Yellow Pincushion | Annual herb | Various insects. Bees, moths, butterflies | | Yellow | Jan-Aug | Typically sand or gravel | Fast, Medium | Moderately Easy |
| *Corethrogyne filaginifolia* | Common Sandaster; California Sandaster; Whiteleaf Sandaster | Perennial herb | Birds, a variety of insects - bees, moths, butterflies | | Pink, Purple, White | Jun-Oct | Tolerant of sand and clay | Fast, Medium, Slow | Moderately Easy |
| *Datura wrightii* | Sacred Thorn-Apple; Sacred Datura; Sacred Thorn-apple | Perennial herb | Butterflies. Primarily insects, including sphinx moths and various beetles | | Purple, White | Feb-Oct | Adaptable but prefers coarse, well-drained soil | Fast, Medium | Moderately Easy |

| Scientific | Common | Plant Type | Wildlife Support | Cultural Use | Flowers | Bloom Time | Soil | Drainage | Ease of Care |
|---|---|---|---|---|---|---|---|---|---|
| Dudleya lanceolata | Lanceleaf Liveforever | Perennial herb, Succulent | Hummingbirds, birds, insects, butterflies, moths | | Orange, Pink, Red, Yellow | May-Jul | Tolerates sand and clay but prefers very rocky substrate | Fast, Medium, Slow | |
| Encelia californica | Bush Sunflower; California Brittlebush | Shrub | Numerous insects are attracted to the flowers, including butterflies and bees. Small birds such as goldfinches are attracted to the seed heads. | Medicine. Boiled to treat toothache. | Brown, Purple, Yellow | Feb-Jun | Tolerates a wide variety of soils | Fast, Medium, Slow | Very Easy |
| Encelia farinosa | Brittlebush; Goldenhills; White Brittlebush | Shrub | Many desert birds, small mammals and insects | | Brown, Yellow | Jan-May | Prefers sandy or decomposed granite soil | Fast, Medium | Very Easy |
| Ericameria palmeri | Palmer's Goldenbush | Shrub | Numerous insects are attracted to the flowers in late summer/fall. Numerous seed-eating birds and small mammals are attracted to the seeds. | | Yellow | Sep-Nov | Adaptable | Fast, Medium, Slow | |
| Eriogonum elongatum | Longstem Buckwheat | Perennial herb | Birds, butterflies, bees | | Cream, Pink, White | Aug-Nov | | | Moderately Easy |
| Eriogonum fasciculatum | Eastern Mojave Buckwheat; Flattop Buckwheat; Yellow Buckwheat | Shrub | Bees, butterflies, birds | | Yellow, Cream, Pink, White | Apr-Sep | Prefers loamy soils | Fast, Medium, Slow | Very Easy |
| Eriogonum gracile | Slender Woolly Buckwheat | Annual herb | Birds, native bees, predatory or parasitoid insects, butterflies: Mormon Metalmark, Bramble Hairstreak Butterfly, Comstock's Hairstreak, Bernardino Dotted-Blue, Small Dotted-Blue, Acmon Blue, Lupine Blue | | Cream, Pink, White, Yellow, Brown | May-Sep | | Fast | |
| Eschscholzia californica | California Goldenpoppy; California Poppy | Annual herb, Perennial herb | Birds, small herbivores, butterflies, bees, moths, other pollinators. | | Orange, Yellow | Feb-Sep | Prefers sandy, infertile, well-drained soils. | Fast, Medium, Slow | Very Easy |

| Scientific | Common | Plant Type | Wildlife Support | Cultural Use | Flowers | Bloom Time | Soil | Drainage | Ease of Care |
|---|---|---|---|---|---|---|---|---|---|
| *Gutierrezia californica* | San Joaquin Snakeweed; California Snakeweed | Perennial herb, Annual herb, Shrub | Various insects are attracted to the flowers. Bees, butterflies | | Yellow, Red | May-Oct | Prefers rocky, gravelly or sandy soil, such as decomposed granite | Fast, Medium | |
| *Helianthus gracilentus* | Slender Sunflower | Perennial herb, Annual herb | Numerous insects, seed-eating birds, bees, butterflies, moths | | Yellow | May-Sep | Prefers sand but tolerates garden soil | Fast | |
| *Hesperoyucca whipplei* | Izote De Hoz; Chaparral Yucca | Succulent | Attracts the Yucca Moth, which co-evolved with this plant. Also attracts California Thrashers, other birds. | | Cream, Pink, Purple, White | Apr-Jun | Prefers rocky soils | Fast | Very Easy |
| *Heterotheca grandiflora* | Telegraphweed; Telegraph Weed | Perennial herb, Annual herb | Insects are attracted to the flowers, especially bees and butterflies | | Yellow | Jan-Dec | Variable | Fast, Medium, Slow | |
| *Keckiella antirrhinoides* | Snapdragon Penstemon; Chaparral Bush-Beard-tongue | Shrub | Hummingbirds, butterflies, bees | | Yellow | Apr-May | Tolerates a variety of soils as long as adequate drainage is provided | Fast, Medium | Moderately Easy |
| *Lasthenia californica* | California Goldfields; Goldfields | Annual herb | Numerous insects, including bees and butterflies, are attracted to the flowers | | Yellow | Feb-Jun | Variable | Medium | Moderately Easy |
| *Layia platyglossa* | Coastal Tidytips | Annual herb | The flowers attract many species of insects, especially butterflies. It is an important nectar plant for Checkerspot butterflies. | | Yellow | Feb-May | Prefers clay or loamy soil, tolerates sandy soil | Fast, Medium, Slow | Moderately Easy |
| *Lupinus sparsiflorus* | Coulter's Lupine; Mojave Lupine | Annual Herb | Various insects are attracted to the flowers. The *Lupinus* genus is host plant to the Arrowhead Blue butterfly. | | Blue, Purple | Mar-Apr | Typically sandy or decomposed granite | Fast | |
| *Malacothamnus fasciculatus* | Mendocino Bushmallow | Shrub | Very attractive to butterflies and small birds, hummingbirds | | Pink | Apr-Jul | Tolerant of a variety of soils as long as drainage is good | Fast, Medium, Slow | Moderately Easy |
| *Marah macrocarpa* | Cucamonga Manroot | Vine, Perennial herb | Butterflies, moths | | White | Jan-Apr | | Fast | |

| Scientific | Common | Plant Type | Wildlife Support | Cultural Use | Flowers | Bloom Time | Soil | Drainage | Ease of Care |
|---|---|---|---|---|---|---|---|---|---|
| *Mirabilis laevis* | Desert Wishbone-Bush | Perennial herb | Butterflies, moths | | Lavender, Purple, White | Feb-May | Variable soil depending on the location and setting | Medium | Moderately Easy |
| *Nemophila menziesii* | Baby Blue-Eyes; Baby Blue Eyes | Annual herb | Numerous insects including butterflies are attracted to the flowers | | Blue | Mar-Jun | Sandy to loamy | Fast, Medium | Moderately Easy |
| *Nicotiana quadrivalvis* | Indian Tobacco | Annual herb | Butterflies, moths | | White, Green, Purple | May-Oct | | | |
| *Penstemon spectabilis* | Showy Penstemon | Perennial herb | Hummingbirds and other birds, butterflies, and bees | | Blue, Pink, Purple | Apr-Jun | Performs best and lives longest in well drained soil | Fast | Moderately Easy |
| *Phacelia ramosissima* | Branched Scorpion-Weed; Branching Phacelia | Annual herb, Perennial herb | Butterflies and bees | | Lavender, White | May-Aug | Prefers sand or sandstone | Fast, Standing | Moderately Easy |
| *Plantago erecta* | Dotseed Plantain | Annual herb | Numerous butterflies, moths. A primary host plant for the federally endangered Quino checkerspot butterfly. | | Brown, White | Spring | Adaptable | Fast, Medium, Slow | |
| *Platystemon californicus* | California Creamcups; Creamcups | Annual herb | Insects, including butterflies and bees. Larval food plant for *Adela oplerella*. | | Cream, White, Yellow | Feb-May | Prefers sandy, gravelly soil. No clay | Fast | |
| *Poa secunda* | One Sided Blue Grass, Pine Bluegrass, Sandberg Bluegrass | Grasses | Butterflies | | Yellow | | Prefers sandy or loamy soils. Does not grow well in clay soils. | Fast, Medium, Slow | Moderately Easy |
| *Ribes divaricatum* | Straggly Gooseberry; Spreading Gooseberry | Shrub | Native bees; Hummingbirds; Butterflies: Tailed Copper, Hoary Comma, Oreas Comma | Food, these wild berries were gathered early summer to late. | Red, Pink, Purple, Green | Mar-May | Moisture retentive but well-drained loamy soil of at least moderate quality | Fast | Moderately Easy |

| Scientific | Common | Plant Type | Wildlife Support | Cultural Use | Flowers | Bloom Time | Soil | Drainage | Ease of Care |
|---|---|---|---|---|---|---|---|---|---|
| *Salvia apiana* | White Sage | Shrub | Hummingbirds, insects, especially carpenter bees and bumble bees | Mid spring flowers for food; late summer seed for food; leaves have medicinal properties, boiled and inhaled for nasal problems, colds and flu; dried plant used for smudging | White | Apr-Jul | Adaptable to a variety of soil types | Fast, Medium | Very Easy |
| *Salvia columbariae* | Chia | Annual herb | Birds, hummingbirds, bees, butterflies, moths | Used for food | Blue, Purple | Mar-Jun | Prefers sandy, well drained soil but tolerates clay | Fast, Medium, Slow | Moderately Easy |
| *Salvia mellifera* | Black Sage | Shrub | Insects, especially bees and butterflies, and hummingbirds are attracted to the flowers. Quail, Towhees and other birds are attracted to the seeds. | Used for food | Blue, Lavender, White | Mar-Jul | Tolerates a variety of soils although prefers good drainage | Fast, Medium | Very Easy |
| *Scrophularia californica* | California Figwort | Perennial herb | Attracts bees, hummingbirds, and a species of small wasp. Figwort is a host plant for the butterfly larvae of Common Buckeye. | | Red | Feb-May | Adaptable to garden soils | Fast, Medium | Very Easy |
| *Silene laciniata* | Mexican Pink; Mexican Catchfly; Mexican Campion; Cardinal Catchfly | Annual herb, Perennial herb | Hummingbirds, butterflies, and moths | | Red | Apr-Jul | Prefers good drainage | Fast, Medium | Moderately Easy |
| *Solanum umbelliferum* | Bluewitch Nightshade; Bluewitch | Shrub | Birds, bees and butterflies | | Blue, Lavender, Purple, Yellow | Jan-Jun | Tolerates a variety of soils | Fast, Medium | Moderately Easy |
| *Solanum xanti* | Chaparral Nightshade; Purple Nightshade | Shrub, Perennial herb | Birds, butterflies, moths | | Blue, Purple | Feb-Jul | Tolerates many soils, sandy, loamy or clay | Fast, Medium, Slow | Moderately Easy |

# Forbland plant palette

| Scientific | Common | Plant Type | Wildlife Support | Cultural Use | Flowers | Bloom Time | Soil | Drainage | Ease of Care |
|---|---|---|---|---|---|---|---|---|---|
| *Artemisia palmeri* | San Diego Sagewort | Shrub | Numerous birds including quail, thrashers, towhees and finches | | Cream, White, Yellow | May-Sep | Typically sandstone or sandy soil | Fast, Medium | Moderately Easy |
| *Asclepias eriocarpa* | Woollypod Milkweed; Kotolo; Indian Milkweed | Perennial herb | Many insects, especially butterflies. important host plant for Monarch butterflies. | Used for materials | Cream, Pink, White | Jun-Aug | Tolerates a variety of soils including clay | Fast, Medium, Slow | Moderately Easy |
| *Castilleja exserta* | Indian Paintbrush; Exserted Indian Paintbrush | Annual Herb | The flowers attract bees and butterflies. This is a crucial host plant for the Bay Checkerspot butterfly, a threatened species in California. The *Castilleja* genus is beneficial for the Leanira Checkerspot and Chalcedon Checkerspot butterflies. | | Lavender, Pink, Purple | Mar-Jun | Tolerant of sand and clay | Fast, Medium, Slow | Moderately Easy |
| *Chaenactis glabriuscula* | Yellow Pincushion | Annual herb | Various insects. Bees, moths, butterflies | | Yellow | Jan-Aug | Typically sand or gravel | Fast, Medium | Moderately Easy |
| *Clarkia purpurea* | Winecup Clarkia; Winecup Fairyfan | Annual herb | Bees, moths, butterflies | | Lavender, Pink, Purple, Red | Apr-Jul | Adaptable | Fast, Medium, Slow | Moderately Easy |
| *Deinandra fasciculata* | Clustered Tarweed | Annual herb | Bees, moths, butterflies | | Yellow | Mar-Oct | Tolerates a variety of soils as long as drainage is good | Fast, Medium | |
| *Ericameria palmeri* | Palmer's Goldenbush | Shrub | Numerous insects are attracted to the flowers in late summer/fall. Numerous seed-eating birds and small mammals are attracted to the seeds. | | Yellow | Sep-Nov | Adaptable | Fast, Medium, Slow | |
| *Eriogonum fasciculatum* | Eastern Mojave Buckwheat; Flattop Buckwheat; Yellow Buckwheat | Shrub | Bees, butterflies, birds | | Yellow, Cream, Pink, White | Apr-Sep | Prefers loamy soils | Fast, Medium, Slow | Very Easy |
| *Eriophyllum confertiflorum* | Golden-Yarrow; Yellow-Yarrow | Annual herb, Perennial herb, Shrub | Very attractive to pollinators, especially butterflies | | Yellow | Feb-Aug | Tolerates clay soil | Medium, Slow | Moderately Easy |

| Scientific | Common | Plant Type | Wildlife Support | Cultural Use | Flowers | Bloom Time | Soil | Drainage | Ease of Care |
|---|---|---|---|---|---|---|---|---|---|
| *Eschscholzia caespitosa* | Tufted Poppy | Annual herb | Birds, bees, butterflies, moths | | Yellow, Orange | Mar-Aug | Prefers average to rich soil | Medium | Moderately Easy |
| *Eschscholzia californica* | California Goldenpoppy; California Poppy | Annual herb, Perennial herb | Birds, small herbivores, butterflies, bees, other pollinators. | Used medicinally | Orange, Yellow | Feb-Sep | Prefers sandy, infertile, well-drained soils. | Fast, Medium, Slow | Very Easy |
| *Koeleria macrantha* | Junegrass; Prairie Junegrass | Grasses | Plants in the *Koeleria* genus are host plant for the Columbian Skipper butterfly | | Yellow, Brown | | Tolerates a variety of soils | Fast, Medium, Slow | Very Easy |
| *Layia platyglossa* | Coastal Tidytips | Annual herb | The flowers attract many species of insects, especially butterflies. It is an important nectar plant for Checkerspot butterflies. | | Yellow | Feb-May | Prefers clay or loamy soil, tolerates sandy soil | Fast, Medium, Slow | Moderately Easy |
| *Lupinus bicolor* | Miniature Lupine; Bicolor Lupine | Perennial herb, Annual herb | Birds, butterflies, bees. The flowers attract numerous insects. Lupines generally are host plant for the Arrowhead Blue butterfly. | | Blue, Lavender, Purple, White | Mar-Jun | Tolerates a variety of soils including very poor soil | Medium | |
| *Lupinus succulentus* | Hollowleaf Annual Lupine; Bigleaf Lupine | Annual Herb | Birds and butterflies. Very attractive to bees. | | Blue, Lavender, White | Feb-May | Tolerates a variety of soils but performs best in heavy, moist soil | Medium, Slow | Moderately Easy |
| *Malacothamnus fasciculatus* | Mendocino Bushmallow | Shrub | Very attractive to butterflies and small birds, hummingbirds | | Pink | Apr-Jul | Tolerant of a variety of soils as long as drainage is good | Fast, Medium, Slow | Moderately Easy |
| *Melica imperfecta* | Smallflower Melicgrass | Grass | Butterflies | | Yellow, Brown | | Tolerates a variety of soils | Fast, Medium | Very Easy |
| *Nasella lepida* | Small Flowered Needlegrass | Grasses | Butterflies and moths | | | Mar-May | Adaptable but often found in clay loam | Medium, Slow | Very Low |
| *Plantago erecta* | Dotseed Plantain | Annual herb | Numerous butterflies, moths. A primary host plant for the federally endangered Quino checkerspot butterfly. | | Brown, White | Spring | Adaptable | Fast, Medium, Slow | |
| *Sidalcea malviflora* | Dwarf Checkerbloom | Perennial herb | Butterflies, native bees, other pollinators | | Pink, Green | May-Aug | Tolerates wide variety of soils | Fast, Medium, Slow | |

| Scientific | Common | Plant Type | Wildlife Support | Cultural Use | Flowers | Bloom Time | Soil | Drainage | Ease of Care |
|---|---|---|---|---|---|---|---|---|---|
| *Solidago velutina* | Threenerve Goldenrod; Sparse Goldenrod; Three-Nerve Goldenrod | Perennial herb | Host plant to the Northern Checkerspot butterfly, and a nectar plant for Monarchs and Skippers, as well as many other pollinators. | | Yellow | Aug-Oct | Tolerates a variety of soils | Medium | Very Easy |
| *Viola pedunculata* | Johnny-Jump-Up | Perennial herb | Butterflies: Variegated Fritillary, Unsilvered Fritillary, Atlantis Fritillary, Callippe Fritillary, Coronis Fritillary, Great Basin Fritillary, Hydaspe Fritillary, Mormon Fritillary, Zerene Fritillary | | Yellow | Feb-Apr | Likes rich soil and no water in summer | Medium, Slow | |

## Alluvial scrub plant palette

| Scientific | Common | Plant Type | Wildlife Support | Cultural Use | Flowers | Bloom Time | Soil | Drainage | Ease of Care |
|---|---|---|---|---|---|---|---|---|---|
| *Ambrosia psilostachya* | Cuman Ragweed; Western Ragweed; Perennial Ragweed | Perennial herb | Birds, butterflies, grasshoppers | | Green | Jul-Nov | Adaptable | Fast, Medium, Slow | |
| *Anemopsis californica* | Yerba-Mansa; Yerba Mansa | Perennial herb | | Used medicinally | White, Cream, Red | Feb-Mar | Tolerant of almost any soil as long as it remains constantly moist | Medium, Slow, Standing | Very Easy |
| *Artemisia californica* | California Sagebrush | Shrub | California Gnatcatcher, Quail, various other birds, insects, butterflies | used medicinally; used for smudging | Cream, White, Yellow | Apr-Oct | Usually found on very dry slopes or sandy soil with low nutrient content, although it is also said to tolerate clay | Fast, Medium, Slow | Very Easy |
| *Artemisia douglasiana* | Douglas' Sagewort; Douglas' Mugwort | Perennial herb | Bees, butterflies, birds | Used medicinally | Cream, White, Yellow | May-Oct | Tolerant of a variety of soils as long as adequate moisture is available | Fast, Medium, Slow | Very Easy |
| *Baccharis salicifolia* | Mule's Fat; Seep Willow; Seepwillow; Seepwillow Baccharis; Mule-Fat | Shrub | Important butterfly and bee plant. Also attracts other beneficial insects | Used medicinally | Pink, White, Yellow | Jan-Dec | Heavier riparian soils, sandy washes | Fast, Medium, Slow, Standing | Very Easy |

| Scientific | Common | Plant Type | Wildlife Support | Cultural Use | Flowers | Bloom Time | Soil | Drainage | Ease of Care |
|---|---|---|---|---|---|---|---|---|---|
| *Croton californicus* | California Croton | Perennial herb | Butterfly and moths | Medicine for earache. Shaman medicine in which care should be used when giving as medicine. | Green | Apr-Jul | Prefers sand or decomposed granite | Fast | Moderately Easy |
| *Cucurbita foetidissima* | Calabazilla; Wild Gourd; Wild Pumkin; Buffalo Gourd; Buffalogourd Pumpkin; Missouri Gourd | Perennial herb, Annual herb | Butterfly and moths | Food / material. Used to wash blankets. Can be irritating to the skin, so rinsing well was important. Seeds can be edible and if necessary they can be gathered and ground up into flour, (a bitter flour) that needs to be leached before eating. Seed can also be boiled first then ground up. | Yellow, Orange | Jun-Aug | Prefers dry sandy or coarse soil | Fast | Moderately Easy |
| *Eriastrum densifolium* | Dense Eriastrum; Giant Woollystar | Perennial herb | Butterflies | | Purple, Lavender | Mar-Sep | | Fast | |
| *Ericameria linearifolia* | Narrowleaf Goldenbush; Slimleaf Goldenbush | Shrub | Birds, butterflies, grasshoppers | | Yellow | Feb-May | | | |
| *Ericameria palmeri* | Palmer's Goldenbush | Shrub | Numerous insects are attracted to the flowers in late summer/fall. Numerous seed-eating birds and small mammals are attracted to the seeds. | | Yellow | Sep-Nov | Adaptable | Fast, Medium, Slow | |
| *Eriogonum fasciculatum* | Eastern Mojave Buckwheat; Flattop Buckwheat; Yellow Buckwheat | Shrub | Bees, butterflies, birds | | Yellow, Cream, Pink, White | Apr-Sep | Prefers loamy soils | Fast, Medium, Slow | Very Easy |

| Scientific | Common | Plant Type | Wildlife Support | Cultural Use | Flowers | Bloom Time | Soil | Drainage | Ease of Care |
|---|---|---|---|---|---|---|---|---|---|
| *Helenium puberulum* | Rosilla | Perennial herb | Butterfly and moth | | Yellow, Cream | Jun-Aug | Prefers sand but tolerates garden soil | Fast, Medium, Standing | |
| *Helianthus annuus* | Common Sunflower; Sunflower; Wild Sunflower; Annual Sunflower | Annual Herb | Sunflowers seeds are very attractive to numerous birds. The flowers are important nectar source for various insects including Monarch and Bordered Patch butterflies. | Food. Seeds gathered and ground into a flour and eaten in different ways. | Yellow, Brown, Orange | Jun-Aug | Adaptable, tolerant of sand, loam and clay | Fast, Medium, Slow | Moderately Easy |
| *Hesperoyucca whipplei* | Izote De Hoz; Chaparral Yucca | Succulent | Attracts the Yucca moth, which co-evolved with this plant. Also attracts California Thrashers and other birds. | | Cream, Pink, Purple, White | Apr-Jun | Prefers rocky soils | Fast | Very Easy |
| *Heterotheca grandiflora* | Telegraphweed; Telegraph Weed | Perennial herb, Annual herb | Insects are attracted to the flowers, especially bees and butterflies | | Yellow | Jan-Dec | Not particular as to soil | Fast, Medium, Slow | |
| *Lepidospartum squamatum* | California Broomsage; California Scalebroom | Shrub | Butterfly and moths | | Cream | Aug-Nov | Adaptable | Fast, Medium, Slow | Moderately Easy |
| *Malacothamnus fasciculatus* | Mendocino Bushmallow | Shrub | Very attractive to butterflies and small birds, hummingbirds | | Pink | Apr-Jul | Tolerant of a variety of soils as long as drainage is good | Fast, Medium, Slow | Moderately Easy |
| *Oenothera californica* | California Evening Primrose | Perennial herb | Butterfly and moths | | White, Pink | Apr-Jun | Prefers sandy, gravelly soil | Fast | Very Easy |
| *Rhus ovata* | Sugar Sumac | Shrub | Insects are attracted to the flowers. Birds are attracted to the fruits. | Used for food; used as material for baskets | White, Pink | Apr-May | Tolerates a variety of soils | Fast, Medium | Very Easy |
| *Rosa californica* | California Wildrose | Shrub | Bees, butterflies and birds | Food. Flower turns into rose hip, gathered and dried and ground up, or after drying it was soaked in water and eaten with other food. | Red, Pink, White | May-Aug | Tolerates clay but does best in moist loamy soil | Medium, Slow | Very Easy |

| Scientific | Common | Plant Type | Wildlife Support | Cultural Use | Flowers | Bloom Time | Soil | Drainage | Ease of Care |
|---|---|---|---|---|---|---|---|---|---|
| *Salvia apiana* | White Sage | Shrub | Hummingbirds, insects, especially carpenter bees and bumble bees | Mid spring flowers for food; late summer seed for food; leaves have medicinal properties, boiled and inhaled for nasal problem, colds and flu; dried plant used for smudging | White | Apr-Jul | Adaptable to a variety of soil types | Fast, Medium | Very Easy |
| *Senecio flaccidus* | Threadleaf Ragwort; Threadleaf Groundsel | Shrub | Butterfly and moths | | Yellow | Jun-Oct | | Fast | |
| *Thalictrum fendleri* | Fendler's Meadowrue; Fendler's Meadow-Rue | Perennial herb | Butterfly and moths | | Yellow | Apr-Jul | Prefers loamy soil with organic matter | Medium, Slow | Moderately Easy |

## Alkali meadow plant palette

| Scientific | Common | Plant Type | Wildlife Support | Cultural Use | Flowers | Bloom Time | Soil | Drainage | Ease of Care |
|---|---|---|---|---|---|---|---|---|---|
| *Anemopsis californica* | Yerba-Mansa; Yerba Mansa | Perennial herb | | Medicinal | White, Cream, Red | Feb-Mar | Tolerant of almost any soil as long as it remains constantly moist | Medium, Slow, Standing | Very Easy |
| *Atriplex serenana* | Bracted Saltweed; Bractscale | Annual herb | Butterflies, moths | | Green | Apr-Oct | | | |
| *Centromadia pungens* | Pungent False Tarplant | Annual herb | Bees, butterflies | | Yellow | Apr-Sep | | Standing | |
| *Distichlis spicata* | Desert Saltgrass; Marsh Spikegrass; Saltgrass; Inland Saltgrass; Seashore Saltgrass | Grass | Several species of Skipper butterflies use this species as host plant. A number of birds and small mammals also utilize this plant. | | Yellow | | Prefers sand or sandstone | Fast, Medium, Slow, Standing | Moderately Easy |

| Scientific | Common | Plant Type | Wildlife Support | Cultural Use | Flowers | Bloom Time | Soil | Drainage | Ease of Care |
|---|---|---|---|---|---|---|---|---|---|
| *Eustoma exaltatum* | Catchfly Prairie-Gentian; Catchfly Prairiegentian; Catchfly Prairie Gentian | Perennial herb, Annual herb | | | | Jan-Dec | | | |
| *Heliotropium curassavicum* | Quail Plant; Seaside Heliotrope; Salt Heliotrope | Perennial herb | Butterflies, including western pygmy blue (*Brephidium exile*) | | Blue, Lavender, White | May-Jun | Grows in many soil types, often in saline or alkaline soils. | | |
| *Lepidium dictyotum* | Net Pepperweed; Alkali Pepperweed | Annual herb | Butterflies | | White | Mar-May | | | |
| *Muhlenbergia asperifolia* | Alkali Muhly; Scratchgrass | Grasses | Butterflies, moths | | Purple | | Prefers loamy or clay soils. Grows poorly in sandy soils. | | |
| *Oligomeris linifolia* | Linearleaf Combess; Lineleaf Whitepuff | Annual herb | | | White | Feb-Jul | | | |
| *Persicaria lapathifolia* | Curlytop Knotweed; Curltop Ladysthumb; Dock-Leaf Smartweed; Nodding Smartweed; Pale Smartweed; Curlytop Smartweed | Annual herb | Butterflies, moths | | Pink | Aug-Nov | | | |
| *Plantago subnuda* | Tall Coastal Plantain | Perennial herb | Butterflies | | Brown | Apr-Jul | Adaptable | Fast, Medium, Slow | |

| Scientific | Common | Plant Type | Wildlife Support | Cultural Use | Flowers | Bloom Time | Soil | Drainage | Ease of Care |
|---|---|---|---|---|---|---|---|---|---|
| *Ranunculus cymbalaria* | Shore Buttercup; Alkali Buttercup | Perennial Herb | Butterflies and moths | | Yellow | May-Jun | Prefers loamy or clay soils. Grows poorly in sandy soils. | Standing | |
| *Suaeda calceoliformis* | Paiuteweed; Western Seepweed; Pursh Seepweed | Annual herb | Butterflies | | | Jul-Oct | | | |
| *Suaeda nigra* | Alkali Seepweed; Shrubby Seepweed; Torrey's Seepweed; Mojave Seablite | Perennial herb | Butterflies | | Cream | May | Usually found in rocky, sandy or gravelly soil with subterranean water | Fast | Moderately Easy |
| *Symphyotrichum lanceolatum* | White Panicle Aster | Perennial herb | Butterflies | | | Jul-Aug | | | |
| *Urtica dioica* | California Nettle; Slender Nettle; Tall Nettle; Stinging Nettle | Perennial herb | Butterflies, moths | | | May-Sep | | | |

# Riparian forest/scrub plant palette

| Scientific | Common | Plant Type | Wildlife Support | Cultural Use | Flowers | Bloom Time | Soil | Drainage | Ease of Care |
|---|---|---|---|---|---|---|---|---|---|
| *Artemisia douglasiana* | Douglas' Sagewort; Douglas' Mugwort | Perennial herb | Bees, butterflies, birds | Used medicinally | Cream, White, Yellow | May-Oct | Tolerant of a variety of soils as long as adequate moisture is available | Fast, Medium, Slow | Very Easy |
| *Baccharis salicifolia* | Mule's Fat; Seep Willow; Seepwillow; Seepwillow Baccharis; Mule-Fat | Shrub | This is an important butterfly and bee plant. Also attracts other beneficial insects | Used medicinally | Pink, White, Yellow | Jan-Dec | Heavier riparian soils, sandy washes | Fast, Medium, Slow, Standing | Very Easy |
| *Baccharis salicina* | Willow Baccharis | Shrub | Host plant to the Common Buckeye butterfly. Many beneficial insects, birds, bees, butterflies | | Cream, White | Aug-Dec | Tolerates a variety of soils as long as adequate moisture is present | Fast, Medium, Slow | Very Easy |
| *Epipactis gigantea* | Heleborina Gigante; Giant Helleborine; Stream Orchid | Perennial herb | | | Orange, Red, Yellow, Green, Purple, Brown | May-Jul | Tolerant of sand and clay | Fast, Medium, Slow | |
| *Euthamia occidentalis* | Western Goldentop; Western Goldenrod | Perennial herb | Butterfly and moth | | Yellow, Green | Apr-Oct | Tolerant of a variety of garden soils as long as sufficient moisture is available | Fast, Medium, Slow, Standing | |
| *Helenium puberulum* | Rosilla | Perennial herb | Butterfly and moth | | Yellow, Cream | Jun-Aug | Prefers sand but tolerates garden soil | Fast, Medium, Standing | |
| *Helianthus annuus* | Common Sunflower; Sunflower; Wild Sunflower; Annual Sunflower | Annual Herb | Sunflowers seeds are very attractive to numerous birds. Important nectar source for various insects including Monarch and Bordered Patch butterflies. | Food. Seeds gathered and ground into a flour and eaten in different ways. | Yellow, Brown, Orange | Jun-Aug | Adaptable, tolerant of sand, loam and clay | Fast, Medium, Slow | Moderately Easy |

| Scientific | Common | Plant Type | Wildlife Support | Cultural Use | Flowers | Bloom Time | Soil | Drainage | Ease of Care |
|---|---|---|---|---|---|---|---|---|---|
| Juncus xiphioides | Irisleaf Rush | Grasses | Attracts birds, moths | Material, this plant was used to make baskets. Material was gathered and then dried and then dyed and worked in a coiling fashion to create beautiful baskets with different designs. | Brown, Yellow, Red | | Adaptable | Fast, Medium, Slow | Very Easy |
| Lupinus latifolius | Broadleaf Lupine; Broad-Leaf Lupine | Perennial Herb | Butterflies, bees, moths, birds | | Blue, Purple | Apr-May | Typically found on soils that are shallow, coarse-textured, rocky and fast draining. | Fast | moderately easy |
| Oenothera elata | Hooker's Eve Primrose; Hooker's Evening primrose; Western Evening primrose | Perennial herb | Many insects use this plant, particularly the large Sphinx moths. Butterflies, hummingbirds and smaller birds such as Goldfinches. | | Yellow, Orange | Jun-Sep | Tolerates virtually any soil | Medium, Slow | Very Easy |
| Platanus racemosa | California Sycamore | Tree | Important for Western Tiger Swallowtail butterfly and other butterflies, hummingbirds, moths | Material used for construction: buildings and tools. Seed pods ground up and used as itching powder medicine | Yellow, Cream, Orange, Brown | Feb-May | Tolerates sand and clay | Fast, Medium, Slow | Very Easy |
| Populus fremontii | Fremont Cottonwood | Tree | Insects, especially butterflies and birds | Leaves used medicinally; wood used to make musical instruments | White, Cream | Feb-Mar | Accepts either sandy or clay soil as long as there is sufficient water | Fast, Medium, Slow | Very Easy |
| Ribes divaricatum | Straggly Gooseberry; Spreading Gooseberry | Shrub | Native bees, hummingbirds, butterflies: Tailed Copper, Hoary Comma, Oreas Comma | Food, these wild berries were gathered early summer to late. | Red, Pink, Purple, Green | Mar-May | Moisture retentive but well-drained loamy soil of at least moderate quality | Fast | Moderately Easy |

| Scientific | Common | Plant Type | Wildlife Support | Cultural Use | Flowers | Bloom Time | Soil | Drainage | Ease of Care |
|---|---|---|---|---|---|---|---|---|---|
| *Salix exigua* | Desert Willow; Sandbar Willow; Coyote Willow; Narrowleaf Willow | Tree, Shrub | Plants in the genus *Salix* are host to a wide variety of pollinators including the Dreamy Duskywing, Viceroy, Lorquin's Admiral, Wiedemeyer's Admiral, Mourning Cloak, Western Tiger Swallowtail, Sylvan Hairstreak, various moths, and some gall-forming wasps. Birds, such as the Least Bell's Vireo and Southwetern Willow Flycatcher, prefer to nest in large, dense willow thickets. | Material /food, useful in construction of large structures. Pod in early spring that can be eaten in time of famine. | Yellow, White | Feb-Mar | Tolerant of various soils as long as there is abundant moisture available | Slow, Standing | |
| *Salix gooddingii* | Goodding's Black Willow; Goodding's Willow | Tree | Plants in the genus *Salix* are host to a wide variety of pollinators including the Dreamy Duskywing, Viceroy, Lorquin's Admiral, Wiedemeyer's Admiral, Mourning Cloak, Western Tiger Swallowtail, Sylvan Hairstreak, various moths, and some gall-forming wasps. Birds, such as the Least Bell's Vireo and Southwetern Willow Flycatcher, prefer to nest in large, dense willow thickets. | Material / medicine. Has many material uses: construct building, granaries, to store food. Resists bugs. Has some pain-relieving properties. Willow is considered a sacred plant among many native people. | Green | Feb-Mar | Tolerates a variety of soils as long as adequate moisture is present | Slow, Standing | |
| *Salix laevigata* | Red Willow | Tree | Insects, butterflies, birds | Material / medicine. Has many material uses: construct building, granaries, to store food. Resists bugs. Has some pain-relieving properties. Willow is considered a sacred plant among many native people. | Cream, Yellow, Red | Feb-May | Heavy moist soils | Fast, Medium, Slow, Standing | Very Easy |
| *Stachys ajugoides* | Bugle Hedge-nettle | Perennial herb | Hummingbirds, birds, bees, moths | | Pink | Apr-Sep | | | |

| Scientific | Common | Plant Type | Wildlife Support | Cultural Use | Flowers | Bloom Time | Soil | Drainage | Ease of Care |
|---|---|---|---|---|---|---|---|---|---|
| *Stachys albens* | White Hedgenettle; Whitestem Hedgenettle | Perennial herb | Hummingbirds, birds, bees, moths | | Pink, White | Jun-Aug | Tolerant of a variety of garden soils as long as sufficient moisture is available | Fast, Medium, Slow | |
| *Vitis girdiana* | Desert Wild Grape; Southern California Grape; Valley Grape | Shrub, Vine | Numerous birds and mammals are attracted to the fruit | Food. Fruit can be eaten or dried into raisins, or ground into a powder for food. | Green | May-Jun | Tolerates a variety of soils | Medium | Very Easy |

## Freshwater marsh plant palette

| Scientific | Common | Plant Type | Wildlife Support | Cultural Use | Flowers | Bloom Time | Soil | Drainage | Ease of Care |
|---|---|---|---|---|---|---|---|---|---|
| *Bidens laevis* | Smooth Beggarticks; Bur Marigold | Perennial herb | Butterflies and moths | | Yellow | Aug-Sep | Adaptable, tolerant of sand, loam and clay | | |
| *Carex diandra* | Lesser Tussock Sedge; Lesser Panicled Sedge | Grasses | Butterflies and moths | | Green, White, Red, Brown | | Adaptable, tolerant of sand, loam and clay | | |
| *Eleocharis montevidensis* | Sand Spikerush | Grasses | Birds, butterflies, and moths | | Brown | | Grows well in moist, sandy spots | Fast | |
| *Equisetum hyemale* | Horsetail; Scouring Horsetail; Scouringrush; Tall Scouring-Rush; Western Scouringrush | Fern | | | | | Tolerates a variety of soils | Fast, Medium, Slow | |

| Scientific | Common | Plant Type | Wildlife Support | Cultural Use | Flowers | Bloom Time | Soil | Drainage | Ease of Care |
|---|---|---|---|---|---|---|---|---|---|
| *Juncus textilis* | Basket Rush | Grasses | Butterflies and moths | Material, this plant was used to make baskets. Material was gathered and then dried and then dyed and worked in a coiling fashion to create beautiful baskets with different designs. | Green, Brown | | Tolerant of a variety of garden soils as long as sufficient moisture is available | Fast, Medium, Slow | Moderately Easy |
| *Juncus xiphioides* | Irisleaf Rush | Grasses | Butterflies and moths | Material, this plant was used to make baskets. Material was gathered and then dried and then dyed and worked in a coiling fashion to create beautiful baskets with different designs. | Brown, Yellow, Red | | Adaptable | Fast, Medium, Slow | Very Easy |
| *Lycopus americanus* | Waterhorehound; Cut-Leaf Water-Horehound; American Bugleweed; Water Horehound; American Waterhorehound; American Water Horehound | Perennial herb | | | White | Jun-Jul | Prefers loamy or clay soils. Grows poorly in sandy soils. | Standing | |
| *Marsilea vestita* | Hairy Pepperwort; Water Clover; Hairy Waterclover | Fern | | | | | | | Moderately Easy |
| *Nasturtium officinale* | Watercress | Perennial herb | Butterflies and moths | | White, Green | May-Sep | | | |

| Scientific | Common | Plant Type | Wildlife Support | Cultural Use | Flowers | Bloom Time | Soil | Drainage | Ease of Care |
|---|---|---|---|---|---|---|---|---|---|
| *Plantago subnuda* | Tall Coastal Plantain | Perennial herb | Butterflies and moths | | Brown | Apr-Jul | Adaptable | Fast, Medium, Slow | |
| *Pluchea odorata* | Sweetscent; Marsh Fleabane | Perennial herb, Annual herb | Butterflies and moths | | Pink | Jun-Nov | Adaptable | Fast, Medium, Slow | |
| *Ranunculus cymbalaria* | Shore Buttercup; Alkali Buttercup | Perennial Herb | Butterflies and moths | | Yellow | May-Jun | Prefers loamy or clay soils. Grows poorly in sandy soils. | Standing | |
| *Schoenoplectus americanus* | Chairmaker's Bulrush; American Bulrush; Olney's Bulrush; Schoenoplectus | Grasses | Butterflies and moths | | Orange, Purple, Red, Brown | | Tolerant of a variety of soils as long as sufficient moisture is available | Fast, Medium, Slow, Standing | |
| *Stachys ajugoides* | Bugle Hedgenettle | Perennial herb | Hummingbirds, birds, bees, butterflies, and moths | | Pink | Apr-Sep | | | |
| *Stachys albens* | White Hedgenettle; Whitestem Hedgenettle | Perennial herb | Hummingbirds, birds, bees, butterflies, and moths | | Pink, White | Jun-Aug | Tolerant of a variety of garden soils as long as sufficient moisture is available | Fast, Medium, Slow | |
| *Trifolium wormskioldii* | Cows Clover; Cow Clover; Sierra Clover | Perennial herb | Bees, butterflies, and moths | | Red, Purple | May-Jun | Prefers loamy soils | Medium, Slow | |

www.ingramcontent.com/pod-product-compliance
Lightning Source LLC
Chambersburg PA
CBRC102014120526
44592CB00031B/93